COMPUTER MODELING
OF CHEMICAL REACTIONS
IN ENZYMES AND SOLUTIONS

COMPUTER MODELING OF CHEMICAL REACTIONS IN ENZYMES AND SOLUTIONS

ARIEH WARSHEL
University of Southern California

A Wiley-Interscience Publication

JOHN WILEY & SONS, INC.

New York   **Chichester**   **Brisbane**   **Toronto**   **Singapore**

In recognition of the importance of preserving what has been
written, it is a policy of John Wiley & Sons, Inc., to have books
of enduring value published in the United States printed on
acid-free paper, and we exert our best efforts to that end.

Library of Congress Cataloging in Publication Data:

Warshel, Arieh.
    Computer modeling of chemical reactions in enzymes and solutions /
    Arieh Warshel.
        p.   cm.
    "A Wiley-Interscience publication."
    Includes index.
    1. Enzymes--Computer simulation.   2. Chemical reactions--Computer
    simulation.   I. Title.
    QP601.W357   1991
    547.7'58--dc20   90-29292
        ISBN 0-471-53395-5   CIP

Printed in the United States of America

10 9 8 7 6 5 4 3 2 1

To My Wife and Parents

CONTENTS

9. HOW DO ENZYMES REALLY WORK?   208

INDEX   229

PREFACE

Most chemical processes in biological systems are performed and accelerated by enzymes. Understanding the action of enzymes is both a challenge
and a fundamental requirement for further quantitative understanding of
biochemical processes. In recent years, enormous progress has been made in
this direction due, in part, to the accumulation of three dimensional
structures of enzymes and the advance of genetic engineering. Yet, despite
these developments, it is not entirely clear how to relate the structures of
enzymes to their catalytic activities or how to examine the feasibility of
different catalytic mechanisms in a well-defined way. Probably, these important issues could not have been addressed on a quantitative level before
the emergence of computer modeling approaches.

Written with these issues in mind this book attempts to simplify the key
concepts of chemical bonding so that they can be used to correlate the
structure and function of proteins. The reader is encouraged to define
enzyme mechanisms in terms of well-defined physical statements about
bonding and charge distribution that can then be analyzed by computer
simulation approaches. The general philosophy of the book reflects the
opinion that, while we may still be arguing about the way different enzymes
work, we have clearly reached the stage where key mechanistic problems
should be addressed in terms of the underlying energetics, and where some
incorrect mechanisms may be excluded.

The book can be used in a one semester course for senior undergraduate
and graduate students who are interested in understanding physical aspects
of biochemistry and computer modeling of macromolecules. It can also be

used as a self-study text and as a complement to other books. Although it follows a rigorous introductory outline, this book does not require significant prior knowledge, and many of the principles taught in the first three chapters can be adopted as working recipes even without a full understanding of their exact derivations. Several specific examples of enzymatic reactions are presented and analyzed to illustrate the approaches needed for simulating such reactions. These examples can be followed conveniently in studies of other systems. Many problems and computer exercises are provided to help readers test their understanding of actual modeling concepts and prepare them to handle larger molecular simulation packages in studies of biophysical problems. This includes the package, ENZYMIX, which follows the formulation of the book quite closely.

I have tried to address, in addition to the technological issues of simulating enzyme functions, the recent scientific revolution with regard to the role of theoretical approaches in biophysics. Traditionally, one formulated an hypothesis to explain the action of a biological molecule and then attempted to examine this hypothesis by some set of experiments. However, biological systems are very complex and in many cases it has been impossible to distinguish between different hypotheses through experimental studies. Apparently, the functions of biological molecules are determined by many interactions which are hard to dissect experimentally. However, computer modeling approaches provide an ideal way of counting and assessing the *totality* of the interactions between the protein and its cofactors. Thus one can now use computers to examine directly a detailed hypothesis and then use experiments to indirectly verify the reliability of the given computer model (rather than to examine the given hypothesis). For example, a computer model that is being used to examine the contribution of strain energy to enzyme catalysis cannot be verified by comparing the calculated and observed strain effect, since we do not know how to evaluate experimentally the magnitude of this effect. Yet, the reliability of such a computer model can be assessed by examining its ability to reproduce the observed effect of mutations on the structure of the protein. With this perspective in mind, I have tried to encourage the readers to view computer simulation methods as "experimental" approaches which are designed to examine different hypotheses about the function of biological molecules. Such approaches may help in turning many concepts in biophysics and biochemistry from qualitative classifications of different effects to quantitative statements about how biological molecules really work.

I thank Shaohua Wang, Greg King, N. Vaidehi and Steve Creighton for helping with the manuscript and with the computer programs. I am also grateful to Henry Chu and Luigi Manna for assisting with the illustrations and to Attila Szabo for his insightful comments on the manuscript.

ARIEH WARSHEL

Los Angeles, California
August 1991

COMPUTER MODELING
OF CHEMICAL REACTIONS
IN ENZYMES AND SOLUTIONS

1

BASIC PRINCIPLES OF CHEMICAL BONDING

All molecules, small and large, are built from atoms which are connected to each other by chemical bonds. The chemistry of any molecule is determined by the type of bonds that exist between its atoms. Thus, to understand any molecular process, it is first necessary to have a good background in the basic principles of chemical bonding. This chapter provides such a background by developing a simple picture of chemical bonding and introducing the concept of molecular *potential surfaces*.

After a brief consideration of the *molecular orbital* (MO) theory of chemical bonding we will spend the better part of the chapter developing the *valence bond* (VB) theory, emphasizing the semi-empirical VB methodology. Although the VB theory is difficult to implement *rigorously* on computers, it can provide a very good understanding of the basics of chemical bonding in organic molecules. With an understanding of some simple aspects of the VB theory, the reader will be able to tackle the chemical aspects of enzyme catalysis.

1.1. THE ISOLATED ATOM

The atom is the building block of all molecules. When an atom is incorporated into a molecule it still retains much of its identity. We will begin our

1

study of chemical bonds by considering the properties of electrons in individual isolated atoms. The best place to start is the simplest system—the one-electron atom.

The hydrogen atom is composed of a negatively charged electron orbiting a positively charged (and much more massive) nucleus. The probability of finding the electron at any given point around the nucleus can be described by functions which are called *atomic orbitals*. These functions tell us where the electron is most likely to be found and thus provide important insight into the nature of chemical bonding between different atoms. Some mathematical background about the concept of atomic orbitals is given below.

1.1.1. The Schroedinger Equation

The behavior of a single electron in an isolated atom can be exactly determined (neglecting relativity) by solving the Schroedinger equation

$$\mathbf{H}(\mathbf{r})\Psi(\mathbf{r}) = (\mathbf{T} + \mathbf{V})\Psi(\mathbf{r}) = E\Psi(\mathbf{r}) \tag{1.1}$$

where $\mathbf{r}$ is the position vector of the electron. $\mathbf{H}$ is the *Hamiltonian* operator, which is given as the sum of the kinetic energy operator $\mathbf{T}$, and the potential energy $\mathbf{V}$. E is the total energy of the system and the function Ψ, which is called the *wave function*, describes the probability distribution for the electron. That is, the probability of finding the electron in a volume element of dimensions $dx\,dy\,dz$ located around the point (x, y, z) in space, is given by $\Psi(x, y, z)^{*}\Psi(x, y, z)\,dx\,dy\,dz$.

The Hamiltonian for a single electron in orbit around a fixed nucleus of charge Z is

$$\mathbf{H} = -\frac{h^2}{8\pi^2 m}\nabla^2 - \frac{e^2 Z}{r} \tag{1.2}$$

where ∇^2 is the Laplacian operator given by $\partial^2/\partial x^2 + \partial^2/\partial y^2 + \partial^2/\partial z^2$, m and e are, respectively, the mass and charge of the electron, h is Planck's constant and Ze is the nuclear charge. Converting the Hamiltonian to *atomic units* (au) we obtain the following:

$$\left(-\frac{1}{2}\nabla^2 - \frac{Z}{r}\right)\Psi(r) = E\Psi(r) \tag{1.3}$$

where in au ($e = h/2\pi = m = 1$). This equation can be solved exactly and the corresponding set of solutions are called the *hydrogen-like* atomic orbitals. These wave functions can be used to describe any one-electron atom (or ion). The hydrogen-like orbitals depend on three quantum numbers (n, l, and m) and their functional forms are tabulated in many books (e.g., Refs. 1 and 2). Here, we only list the two lowest energy orbitals using the notation $\chi_{n,l,m}$ for the wave functions.

$$\chi_{100} = \pi^{-1/2} Z^{3/2} \exp[-Zr]$$

$$\chi_{200} = (32\pi)^{-1/2} Z^{3/2} (2 - Zr) \exp[-Zr/2] \tag{1.4}$$

The energies associated with these atomic orbitals are given by

$$E_n = -\frac{Z^2}{2n^2} \tag{1.5}$$

Exercise 1.1. Verify eq. (1.5) using eq. (1.3) and χ_{100} of eq. (1.4). Hint: Use the ∇^2 operator in the spherical coordinates representation

$$\nabla^2(\theta, \phi, r) = \frac{1}{r^2} \frac{\partial}{\partial r} \left(r^2 \frac{\partial}{\partial r} \right) + \frac{1}{r^2 \sin \theta} \frac{\partial}{\partial \theta} \left(\sin \theta \frac{\partial}{\partial \theta} \right) + \frac{1}{r^2 \sin^2 \theta} \frac{\partial^2}{\partial \phi^2}$$

Solution 1.1. Since χ is independent of θ and ϕ, we can write $\nabla^2 \chi(r) = \frac{1}{r^2} \frac{\partial}{\partial r} (r^2 \frac{\partial}{\partial r}) \chi(r)$. Substituting in eq. (1.3) we obtain $\frac{Z}{r} \chi(r) - \frac{Z^2}{2} \chi(r) - \frac{Z}{r} \chi(r) = E\chi(r)$, and $E_1 = -\frac{Z^2}{2}$.

1.1.2. Wave Functions for Atoms

To describe atoms with several electrons, one has to consider the interaction between the electrons, adding to the Hamiltonian a term of the form $\Sigma_{i<j} \frac{1}{r_{ij}}$. Despite this complication it is common to use an approximate wave function which is a product of hydrogen-like atomic or'.tals. This is done by taking the orbitals in order of increasing energy and assigning no more than two electrons per orbital.

The wave function, constructed from the atomic orbitals must be antisymmetric with respect to interchange of electrons in order to satisfy the *Pauli exclusion principle*, having different spin quantum numbers (α and β) for two electrons which are in the same orbital.

For example, the helium atom electron wave function can be written as

$$\Psi_{\text{He}} = \frac{1}{2^{1/2}} (\chi_{100}(1)\alpha(1)\chi_{100}(2)\beta(2) - \chi_{100}(1)\beta(1)\chi_{100}(2)\alpha(2)) \tag{1.6}$$

This type of wave function, which is clearly antisymmetric with respect to exchange of electron 1 and 2, can be also written in a determinant form

$$\frac{1}{\sqrt{2}} \begin{vmatrix} \chi_{100}(1)\alpha(1) & \chi_{100}(2)\alpha(2) \\ \chi_{100}(1)\beta(1) & \chi_{100}(2)\beta(2) \end{vmatrix} \tag{1.7}$$

Such a wave function is known as a *Slater determinant*. In general, when we deal with antisymmetrized wave functions, we use a compact notation for the Slater determinant:

$$\frac{1}{2^{1/2}} \left(\chi_A(1)\alpha(1)\chi_B(2)\beta(2) - \chi_A(1)\beta(1)\chi_B(2)\alpha(2) \right) = |\chi_A \bar{\chi}_B| \quad (1.8)$$

where the vertical bar implies determinant and the horizontal bar indicates that the corresponding wave function is associated with a spin function β.

1.1.3. Valence Electrons and the Core/Valence Separation

When multi-electron atoms are combined to form a chemical bond they do not utilize all of their electrons. In general, one can separate the electrons of a given atom into inner-shell core electrons and the valence electrons which are available for chemical bonding. For example, the carbon atom has six electrons, two occupy the inner 1s orbital, while the remaining four occupy the 2s and three 2p orbitals. These four can participate in the formation of chemical bonds. It is common practice in semi-empirical quantum mechanics to consider only the outer valence electrons and orbitals in the calculations and to replace the inner electrons + nuclear core with a screened nuclear charge. Thus, for carbon, we would only consider the 2s and 2p orbitals and the four electrons that occupy them and the +6 nuclear charge would be replaced with a +4 screened nuclear charge.

1.2. MOLECULAR ORBITALS FOR DIATOMIC MOLECULES

The atomic orbitals considered above can be used to help describe the wave functions of electrons in chemical bonds. To see this, we start with the simple problem of the H_2^+ molecule (Fig. 1.1).

The Hamiltonian for this system should include the kinetic and potential energy of the electron and both of the nuclei. However, since the electron mass is more than a thousand times smaller than that of the lightest nucleus, one can consider the nuclei to be effectively motionless relative to the quickly moving electron. This assumption, which is basically the *Born–Oppenheimer approximation*, allows one to write the Schroedinger equation neglecting the nuclear kinetic energy. For the H_2^+ ion the Born–Oppenheimer Hamiltonian is

$$\mathbf{H}\psi(R, r_A, r_B) = \left(-\frac{\nabla^2}{2} - \frac{1}{r_A} - \frac{1}{r_B} + \frac{1}{R} \right) \psi(R, r_A, r_B) = \varepsilon(R)\psi(R, r_A, r_B)$$
$$(1.9)$$

where R is the distance between the two nuclei and r_A and r_B are the distances of the electron from nucleus A and B respectively.

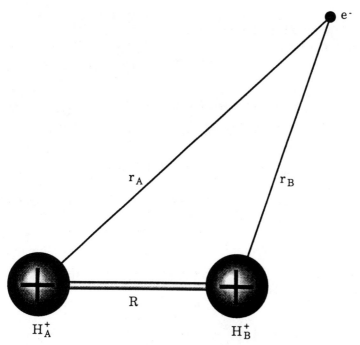

FIGURE 1.1. The H_2^+ molecule.

1.2.1. The MO Potential Surface for the H_2^+ Molecule

In order to obtain an approximate solution to eq. (1.9) we can take advantage of the fact that for large R and small r_A, one basically deals with a hydrogen atom perturbed by a bare nucleus. This situation can be described by the hydrogen-like atomic orbital χ_{100} located on atom A. Similarly, the case with large R and small r_B can be described by χ_{100} on atom B. Thus it is reasonable to choose a linear combination of the atomic orbitals χ_{100}^A and χ_{100}^B as our approximate wave function. Such a combination is called a *molecular orbital* (MO) and is written as

$$\psi = C_A \chi_{100}^A + C_B \chi_{100}^B \tag{1.10}$$

To find the optimal coefficients C_A and C_B one can use the *variation principle*, which states that any trial solution for the wave function will give a larger value for $\varepsilon(R)$ than the value obtained with the exact wave function. With this in mind, we should try to find the minimum of $\varepsilon(R)$ as a function of C_A and C_B. This is done by expressing $\varepsilon(R)$ as

$$\varepsilon(R) = \frac{\int \psi^* H \psi \, d\tau}{\int \psi^* \psi \, d\tau} = \frac{C_A^2 H_{AA} + C_B^2 H_{BB} + 2 C_A C_B H_{AB}}{C_A^2 + C_B^2 + 2 C_A C_B S_{AB}} \tag{1.11}$$

where $H_{AB} = \int \chi_A^* \mathbf{H} \chi_B \, d\tau$ and $S_{AB} = \int \chi_A^* \chi_B \, d\tau$. The optimal Cs are found by solving the set of equations

$$\frac{\partial \varepsilon(R)}{\partial C_A} = \frac{\partial \varepsilon(R)}{\partial C_B} = 0 \tag{1.12}$$

This leads to the well-known *secular equation*

$$C_A(H_{AA} - \varepsilon) + C_B(H_{AB} - S_{AB}\varepsilon) = 0$$

$$C_A(H_{AB} - \varepsilon S_{AB}) + C_B(H_{BB} - \varepsilon) = 0 \tag{1.13}$$

The permissible values of ε are determined by solving the equation

$$\begin{vmatrix} H_{AA} - \varepsilon & H_{AB} - \varepsilon S_{AB} \\ H_{AB} - \varepsilon S_{AB} & H_{BB} - \varepsilon \end{vmatrix} = 0 \tag{1.14}$$

The coefficient vector $\mathbf{C}$ is then solved by substituting the value of ε back into the original secular equation.

Since we deal with two identical nuclei $H_{AA} = H_{BB}$ we obtain

$$\varepsilon_{\pm}(R) = \frac{H_{AA} \pm H_{AB}}{1 \pm S_{AB}} \tag{1.15}$$

Evaluating the energy ε for different values of R gives the effective potential for the nuclei in the presence of the electron. This function is called the *Born–Oppenheimer potential surface* or just the *potential surface*. In order to evaluate $\varepsilon(R)$ we have to determine H_{AA}, H_{AB}, and S_{AB}. These quantities, which can be evaluated using elliptical coordinates, are given by

$$H_{AA} = E_{100} + \gamma + R^{-1}$$

$$H_{BB} = H_{AA}$$

$$H_{AB} = SE_{100} + \beta + SR^{-1}$$

$$H_{BA} = H_{AB} \tag{1.16}$$

where

$$\gamma = -\frac{1}{R}\left[1 - \exp[-2R](1 + R)\right]$$

$$\beta = -\exp[-R](1 + R)$$

$$S = \exp[-R]\left(1 + R + \frac{R^3}{3}\right) \tag{1.17}$$

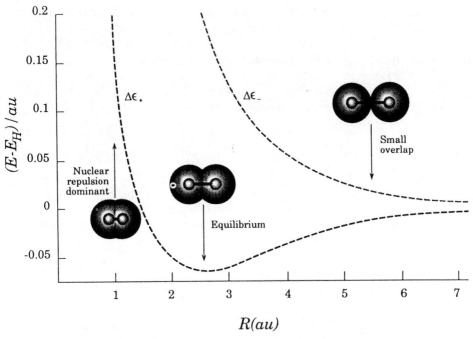

FIGURE 1.2. The potential surfaces of the H_2^+ molecule.

and E_{100} is the energy of an electron in the χ_{100} orbital (which is equal to $-\frac{1}{2}$ au).

The potential surface obtained from eqs. (1.15) and (1.16) is described in Fig. 1.2. Note that at large distances we obtain the energy of the isolated hydrogen atom $(-1/2)$. The coefficients C_A and C_B and the MO wave function is obtained by substituting eq. (1.15) in eq. (1.13) and requiring a *normalized* wave function, for example,

$$\int \psi_+^* \psi_+ \, d\tau = C_A^2 + 2C_A C_B S_{AB} + C_B^2 = 1 \qquad (1.18)$$

The resulting two sets of coefficients (for ε_+ and ε_-) give

$$\psi_\pm = (\chi_A \pm \chi_B)/(2 \pm 2S_{AB})^{1/2} \qquad (1.19)$$

1.2.2. MO Potential Surface for Molecules with Many Valence Electrons

In treating molecules with more than one electron we use *Slater determinants* made of molecular orbitals. For example, for the H_2 molecule we use

$$\Psi_{H_2}^{MO} = |\psi_+ \bar{\psi}_+| \qquad (1.20)$$

The general treatment of $2n$ electrons and n molecular orbitals is outlined in Appendix A and can be skipped by the readers who are not interested in this additional background. Here we only give very simplified versions of this treatment, which will help us in some qualitative discussions.

We start with the oversimplified version of the regular *Huckel* approximation. This approach considers the complete Hamiltonian of a system of $2n$ electrons and m nuclear cores, which is given by

$$\mathbf{H} = \sum_{i}^{2n} \left(-\frac{1}{2} \nabla_i^2 - \sum_{A=1}^{m} \frac{Z_A}{r_{iA}} \right) + \sum_{i<j} \frac{1}{r_{ij}} = \sum_{i=1}^{2n} h(i) + \sum_{i<j} \frac{1}{r_{ij}} \qquad (1.21)$$

This Hamiltonian is then approximated by removing the electron–electron repulsion term and representing its effect by replacing each Z_A by a new effective charge (Z_A') that reflects the screening of the original core charge by the presence of other valence electrons. This gives

$$\mathbf{H} \simeq \sum_{i=1}^{2n} \left(-\frac{1}{2} \nabla_i^2 - \sum_{A=1}^{m} \frac{Z_A'}{r_{iA}} \right) = \sum_{i=1}^{2n} h_{\text{eff}}(i) \qquad (1.22)$$

For this simple Hamiltonian, which involves the sum of one-electron Hamiltonians, we can use a wave function of the form

$$\Psi = \psi_1(1)\psi_1(2)\psi_2(3) \cdots \psi_m(2n) \qquad (1.23)$$

where the ψ_i are the molecular orbitals constructed from the m atomic orbitals by

$$\psi_i = \sum_{\mu}^{m} C_{i\mu} \chi_\mu \qquad (1.24)$$

where μ runs over the atomic orbitals, which are attached to the m cores assuming for simplicity that $m = n$. Here we do not use a Slater determinant wave function but the simple wave function of eq. (1.23), since in the absence of the $1/r_{ij}$ term we get the same energies and MOs for both wave functions. Of course, after solving for the C's and obtaining MO we can take the $2m$ active electrons and place them with the proper spins in a Slater determinant.

Since our Hamiltonian involves a sum of $h_{\text{eff}}(i)$, which are only functions of the coordinates and momenta of a single electron, we can use separation of variables and reduce the problem to m identical one-electron problems

$$h_{\text{eff}}(1)\psi_i(1) = \varepsilon_i \psi_i(1) \qquad (1.25)$$

where the total energy is given by

$$E = \sum_i n_i \varepsilon_i + \sum_{A>B} \frac{Z_A' Z_B'}{R_{AB}} \qquad (1.26)$$

where n_i is the number of electrons in each orbital and where the $Z_A' Z_B'$ term represents the core–core repulsion term.

Equation (1.25) leads to the mth dimensional equivalent of eq. (1.13), which can be written in matrix notation as

$$\mathbf{HC}_i = \varepsilon_i \mathbf{C}_i \qquad (1.27)$$

Here the overlap integrals between different orbitals are neglected ($S_{\mu\nu} = \delta_{\mu\nu}$ where $\delta_{\mu\nu}$ is the *Kronecker delta* which is given by $\delta_{\mu\nu} = 0$ for $\mu \neq \nu$ and $\delta_{\mu\nu} = 1$ for $\mu = \nu$). The elements of the H matrix are given by

$$H_{\mu\mu} = \int \chi_\mu^*(1) h_{\text{eff}}(1) \chi_\mu(1)\, d\tau_1 = \alpha_\mu$$

$$H_{\mu\nu} = \int \chi_\mu^*(1) h_{\text{eff}}(1) \chi_\nu(1)\, d\tau_1 = \beta_{\mu\nu} \qquad (1.28)$$

The energy surface of eq. (1.26) can be written now as

$$E = \sum_i n_i \mathbf{C}_i' \mathbf{HC}_i + \sum_{A>B} Z_a' Z_b' / R_{AB} = \sum_{\mu\nu} P_{\mu\nu} H_{\mu\nu} + \sum_{A>B} Z_a' Z_b' / R_{AB} \qquad (1.29)$$

where the $P_{\mu\nu}$ are called *bond orders* and are given by

$$P_{\mu\nu} = \sum_i n_i C_{i\mu} C_{i\nu} \qquad (1.30)$$

where n_i is the number of electrons in the ith MO (n_i is 1 or 2 for the occupied molecular orbitals and zero for all other orbitals).

The regular Huckel approach presented above does not give correct bonding properties and serves here mainly to give the reader a simple description of a complicated many-electron problem. In order to address realistic problems in our exercises we will move to the simplest approach that gives reasonable bonding properties. This can be obtained by a variant of Appendix A which is called the *iterative extended Huckel method*. In this approach we obtain the same secular equation as in the regular Huckel method, but now the matrix elements of **H** are given by

$$H_{\mu\mu} = \alpha_\mu = \bar{\alpha}_\mu - aQ_\mu - \sum_{B \neq A} 14.4 Q_B (D + R_{AB})^{-1}$$

$$H_{\mu\nu} = \beta_{\mu\nu} = \beta_0 \exp[-\rho R_{\mu\nu}] \qquad (1.31)$$

where from now on we will be using energy units of either electron volt (eV)

(1 au = 27.21 eV) of kcal/mol (1 au = 627.5 kcal/mol) and distance units of Å (1 au = 0.529 Å). The units used in eq. (1.31) and through the rest of this section are eV and Å. In later stages we will switch to kcal/mol and Å.

The parameter $\bar{\alpha}_\mu$ is the ionization energy of an electron from the μth atomic orbital located on the Ath atom and β is the so-called *resonance integral* (represented here by a simple exponential). The Q_B and $P_{\mu\mu}$ terms of α_μ represent corrections to the effective ionization potential due to the residual charges on the different atoms. The charges are determined by

$$Q_A = \sum_{\mu(A)} Q_{\mu\mu} = \sum_{\mu(A)} (Z_\mu - P_{\mu\mu}) = Z_A - P_{AA}$$

$$P_{AA} = \sum_{\mu(A)} P_{\mu\mu} \tag{1.32}$$

Since $P_{\mu\mu}$ depends on the solution of the secular equation, which in turn depends on $P_{\mu\mu}$, it is clear that we must solve iteratively for the molecular orbitals. In general, we will consider only the first few iterations and start the first iteration with $P^0_{\mu\mu} = Z_\mu$, where Z_μ is the effective charge of the nuclear core of the νth orbital (for more than one orbital per atom we have $Z_A = \Sigma_{\mu(A)} Z_\mu$). The potential surface of the system is then approximated by

$$E = \sum_{i=1}^{m} n_i \varepsilon_i + \sum_{A>B} 14.4[Z_A Z_B R_{AB}^{-1} - P_A P_B (D + R_{AB})^{-1}] \tag{1.33}$$

The matrix elements $H_{\mu\mu}$ and $H_{\mu\nu}$ in eq. (1.31) could be evaluated analytically as was done in the simple case of eq. (1.16). However, such a treatment would result in an entirely unreliable potential surface, since the Huckel approximation neglects several important integrals. Instead it is preferable to take a *semi-empirical* approach, representing the integrals β and α by a fairly simple function [as indeed done in eq. (1.31)] with adjustable parameters (β_0 and ρ). These parameters and the parameters $\bar{\alpha}_\mu$ and D can be calibrated by fitting the calculated properties to the corresponding observed properties (e.g., equilibrium structures, bond energies, and excitation energies) for different molecules. In this philosophy it is assumed that the α and the β, which are calibrated on few molecules will give reasonable results for other molecules from the same class.

To clarify this somewhat formal discussion it is important to perform the exercises given below.

Exercise 1.2. Consider a diatomic molecule that has two orbitals (A and B on the atoms A and B, respectively) and two electrons in the MOs formed by these orbitals. What is the Huckel electronic energy of this system in terms of the parameters α_A, α_B, and β_{AB}?

Solution 1.2.   The secular determinant for this problem is:

$$\begin{vmatrix} \alpha_A - \varepsilon & \beta_{AB} \\ \beta_{AB} & \alpha_B - \varepsilon \end{vmatrix} = 0 \tag{1.34}$$

The solution to the 2×2 secular equation given above is

$$\varepsilon_\pm = \frac{1}{2} \left[(\alpha_A + \alpha_B) \pm ((\alpha_A - \alpha_B)^2 + 4\beta^2)^{\frac{1}{2}} \right] \tag{1.35}$$

Exercise 1.3.   Using the equations introduced in the last problems, generate a potential surface for the H_2 molecule assuming that $\bar{\alpha}_H = -13.6 \, \text{eV}$, $\beta = -9.5 \exp[-R] \, \text{eV}$ (with R in Å), $D = 0.3 \, \text{Å}$, and $a = 1$. Consider only the first iteration with $P_H = 1$ and $Z_H = 1$. Adjust the value of β_0 to see how it affects the value of the bond energy (the value of E at its minimum). Such a procedure will clarify the point that β_0 (as well as the other parameters) are chosen semi-empirically to give the best potential surface for our molecule.

Solution 1.3.   Since $\alpha_A = \alpha_B = \alpha$, the expression given above reduces to

$$\varepsilon_\pm = \alpha \pm \beta \tag{1.36}$$

There are two electrons that can both be put into the lower-energy orbital with opposite spins so the electronic energy is $2\varepsilon_+$. The internuclear repulsion term must also be included in the total energy expression, giving (through eq. 1.33):

$$E = 2(-13.6 - 9.5 \exp[-R]) + 14.4R^{-1} - 14.4(D + R)^{-1} \tag{1.37}$$

Exercise 1.4.   Solve the exercise given above for the HF molecule. Consider only the hydrogen 1s orbital ($\bar{\alpha}_H = -13.6 \, \text{eV}$) and one fluorine 2p orbital ($\bar{\alpha}_F = -13.0 \, \text{eV}$). Use the resonance integral $\beta_{HF} = -6 \exp[-2R] \, \text{eV}$, $D = 4 \, \text{Å}$, and $a = 0.5$.

1.2.3.   Solution of Secular Equations by Matrix Diagonalization

The treatment of molecular problems frequently leads to secular equations (e.g., eq. 1.27). Such equations might have much larger dimensions than 2×2, making it difficult and sometimes impossible to solve them analytically. Fortunately, there are numerical methods that allow us to take eq. 1.27 and to find the vectors **C** (which are called *eigenvectors*) and the energies ε (which are called *eigenvalues*). Such methods are called *diagonalization methods* since they transform the matrix **H** into a diagonal matrix by

$$\mathbf{C'HC} = \mathbf{E} \tag{1.38}$$

where **E** is a diagonal matrix with $E_{ii} = \varepsilon_i$ and $E_{ij} = 0$ for $i \neq j$, and **C** is a matrix with the C_i in its columns (e.g., C_1 in the first column). Many diagonalization approaches are now available in standard computer programs which can be treated as black boxes that receive the matrix **H** and give out the eigenvalues ε_i and the eigenvectors C_i. A typical diagonalization subroutine called subroutine DIAG is given in Program 1.A at the end of this chapter and can be used by the reader.

Exercise 1.5. Use subroutine DIAG to evaluate the ε_+ and ε_- for the system of Exercise 1.4.

Solution 1.5. See Program 1.A.

1.2.4. Incorporating the Effect of External Charges in MO Treatments

The approach discussed above can provide a qualitative description of the effect of external fields on bond-breaking processes. For example, consider the H_2 molecule $(H_A - H_B)$ in the presence of an Li^+ ion 3 Å away from H_B on the A–B axis. To study this problem, we assume that there is no charge migration to the Li_C^+ location (so that $P_C = 0$) and that $\beta_{AC} = \beta_{BC} = 0$ since the Li^+ ion is sufficiently far from H_A and H_B. In this case, we can write the **H** matrix as

$$\mathbf{H} = \begin{vmatrix} \alpha_A & \beta_{AB} & 0 \\ \beta_{AB} & \alpha_B & 0 \\ 0 & 0 & \alpha_C \end{vmatrix} \tag{1.39}$$

where we have in the first iteration

$$\alpha_A^0 = \bar{\alpha}_H - a - 14.4(D + R_{AC})^{-1}$$

$$\alpha_B^0 = \bar{\alpha}_H - a - 14.4(D + R_{BC})^{-1}$$

$$\alpha_C^0 = \bar{\alpha}_{Li} \tag{1.40}$$

The index 0 represents the first iteration in our procedure. We can use the subroutine DIAG to solve the secular equation and this is a recommended exercise, but instead we can make use of the fact that the 2×2 matrix of A and B is not mixed with the 1×1 submatrix of C. Thus, we can solve the 2×2 problem, including the effect of the lithium ion, only through its effect on α_A and α_B. This will give the $\varepsilon_\pm$ of eq. (1.35) that upon substitution in eq. (1.27) will give the molecular orbital coefficients (the C's) and then through eq. (1.32) will give the charges Q_A^1 and Q_B^1 of our first iteration (where the superscript one designates the first iteration). Substituting the Q's in the expression for the α's will give α_A^1 and α_B^1 and new energies and new Q's. The calculated energies and wave functions after the first iteration are illustrated schematically in Fig. 1.3.

The main feature of the new wave function is its polarization by the field of the ion. That is, the presence of the ion changes the effective ionization energies of H_A and H_B and tends to pull electrons toward H_B. The effect of the external ion on the potential surface for bond breaking is shown in Fig. 1.4.

Exercise 1.6. Evaluate the potential surface of the $H_2 + Li^+$ system at four points ($R_{AB} = 1$, 2, 3, and 4 Å), using $\bar{\alpha}_H = -13.6$ eV, $a = 1$, $D = 0.3$, and $\beta = -9.5\exp(-R)$ eV.

Solution 1.6. Use Program 1.B, which is given at the end of this chapter.

The Huckel approach does not really consider the interaction between the bonding electrons. Including this electrostatic interaction in the calcula-

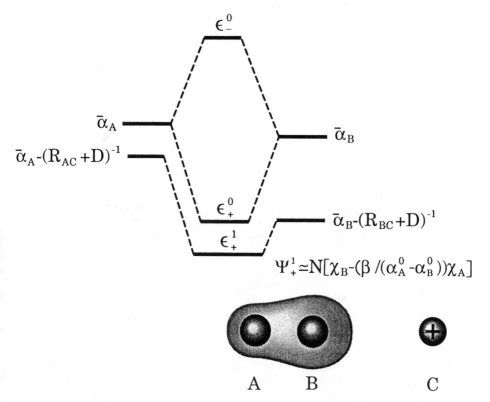

FIGURE 1.3. A schematic illustration of the energies and wave functions of a diatomic molecule in the presence of a positively charged ion. The ground state singlet wave function has a large character of χ_B, since the electron is attracted to the positively charged ion. ε_+^0 and ε_+^1 represent the ground state energy in the first and second iteration, respectively.

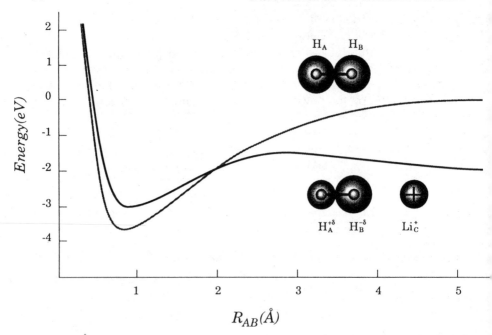

FIGURE 1.4. A simplified MO potential surface for an H_2 molecule in the presence of a Li^+ ion. The Li^+ is held 3 Å from H_B and the bond length R_{AB} is changed by moving H_A.

tions leads to the more complicated self-consistent field approach given in Appendix A. However, even this sophisticated treatment does not give a correct description of the correlated motion of the bonding electrons. Only the configuration interaction (CI) treatment, which is not considered here, can make the MO method give a correct potential surface for a bond-breaking reaction. This is done, however, at the expense of physical insight.

1.3. THE VALENCE BOND DESCRIPTION OF DIATOMIC MOLECULES

The main features of the chemical bonding formed by electron pairs were captured in the early days of quantum mechanics by Heitler and London. Their model, which came to be known as the valence bond (VB) model in its later versions, will serve as our basic tool for developing potential surfaces for molecules undergoing chemical reactions. Here we will review the basic concepts of VB theory and give examples of potential surfaces for bond-breaking processes.

1.3.1. The Heitler–London Treatment of the H_2 Molecule

The original VB wave function was introduced in the treatment of the hydrogen molecule by Heitler and London in 1932. This treatment considered only the one 1s orbital on each hydrogen atom and assumed that the best wave function for a system of two electrons on two different atoms is a product of the two atomic 1s orbitals: $\psi \simeq \chi_{1s}^A \chi_{1s}^B$. This wave function needs to be modified, however, to accommodate the antisymmetry of the wave function and to take into account the spin of the two electrons.

In order to include the spin of the two electrons in the wave function, it is assumed that the spin and spatial parts of the wave function can be separated so that the total wave function is the product of a spin and a spatial wave function: $\Psi \simeq \psi_{space} \Theta_{spin}$. Since our Hamiltonian for the H_2 molecule does not contain any spin-dependent terms, this is a good approximation (NB—the complete Hamiltonian *does* contain spin-dependent terms, but for hydrogen they are rather small and do not appreciably affect the energetics of chemical bonding). For a two-electron system it turns out that there are four possible spin wave functions; they are:

$$\Theta_0^0 = \alpha_1 \beta_2 - \beta_1 \alpha_2$$

$$\Theta_1^1 = \alpha_1 \alpha_2$$

$$\Theta_0^1 = \alpha_1 \beta_2 + \beta_1 \alpha_2$$

$$\Theta_{-1}^1 = \beta_1 \beta_2 \qquad (1.41)$$

where, for example, Θ_0^1 refers to a spin wave function with total spin of 1 and a spin z-component of 0.

Considering first the state with a total spin of 0, we note that since the spin wave function is antisymmetric with respect to interchanging the particle labels, the spatial part of the wave function should be symmetric in order to preserve the overall antisymmetry of the wave function. This leads to the following expression for the wave function:

$$^1\psi_+^{cov} = N[\chi_A(1)\chi_B(2) + \chi_B(1)\chi_A(2)](\alpha_1\beta_2 - \beta_1\alpha_2) \qquad (1.42)$$

where the superscript cov designates a *covalent* bond and N is a normalization constant which is evaluated through the relation

$$\int (^1\psi^{cov})^*(^1\psi^{cov})\, d\tau = 1 \qquad (1.43)$$

For the triplet spin states it is only necessary to consider one of the three possible spin wave functions and we will take the Θ_0^1 wave function. Since the triplet wave function is symmetric with respect to switching the labels of particles 1 and 2, the spatial wave function needs to be antisymmetric. These considerations lead to the following form for the triplet-state wave function:

$$^3\psi_-^{cov} = N^1[\chi_A(1)\chi_B(2) - \chi_B(1)\chi_A(2)](\alpha_1\beta_2 + \beta_1\alpha_2) \qquad (1.44)$$

The H_2 electronic Hamiltonian is

$$H = h(1) + h(2) + 1/r_{12} \qquad (1.45)$$

with the one-electron Hamiltonians:

$$h(1) = \frac{-\nabla_1^2}{2} - r_{1A}^{-1} - r_{1B}^{-1}$$

$$h(2) = \frac{-\nabla_2^2}{2} - r_{2A}^{-1} - r_{2B}^{-1} \qquad (1.46)$$

Evaluating $\langle {}^1\psi^{cov}|H|{}^1\psi^{cov}\rangle$ gives the total ground state energy as

$$^1E_+ = \frac{(J+K)}{(1+S^2)} \qquad (1.47)$$

where

$$J = \langle \chi_A(1)\chi_B(2)|H|\chi_A(1)\chi_B(2)\rangle$$

$$K = \langle \chi_A(1)\chi_B(2)|H|\chi_A(2)\chi_B(1)\rangle$$

$$S^2 = \langle \chi_A(1)\chi_B(2)|\chi_A(2)\chi_B(1)\rangle \qquad (1.48)$$

Similarly, the energy of the triplet state is given by

$$^3E_- = \frac{(J-K)}{(1-S^2)} \qquad (1.49)$$

The integrals J and K are called the *Coulomb* and *exchange* integrals respectively. The actual potential surface of the singlet Heitler–London wave function gives a good description of the real H_2 ground state, including the bond-breaking process (see Fig. 1.5). This is because the wave function tends to keep the electrons apart and mimics the correlation of the electrons. Since at large intranuclear separation the electrons are almost completely localized on different nuclei, the Heitler–London wave function gives the asymptotically correct wave function for bond-breaking processes.

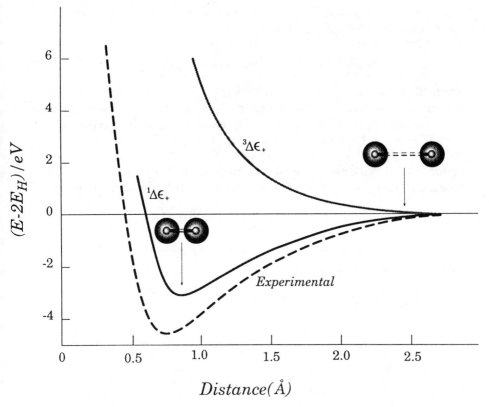

FIGURE 1.5. The VB potential surface of the H_2 molecule and the corresponding experimental potential surface.

1.3.2. Adding Ionic Terms

The Heitler–London wave function given in the last section provides a qualitatively good picture of the bonding in the hydrogen molecule, but has a binding energy that is about 1 eV too small. In order to rectify this problem, it is necessary to consider states that have both the electrons located on one of the hydrogen atoms. These states are called *ionic* states and their addition to the Heitler–London wave function gives the VB theory.

For the H_2 molecule there are two possible ionic states: (H^+H^-) and (H^-H^+). The wave functions for these two ionic states are

$$^1\psi^{\text{ion}}_{H^-H^+} = N^{\text{ion}}\chi_A\chi_A(\alpha_1\beta_2 - \beta_1\alpha_2)$$

$$^1\psi^{\text{ion}}_{H^+H^-} = N^{\text{ion}}\chi_B\chi_B(\alpha_1\beta_2 - \beta_1\alpha_2) \qquad (1.50)$$

Adding the contributions from these ionic wave functions to the ground state of the H_2 molecule gives

$$^1\psi = C_{\text{cov}}\,^1\psi^{\text{cov}} + C_{\text{ion}}(^1\psi^{\text{ion}}_{\text{H}^-\text{H}^+} + {}^1\psi^{\text{ion}}_{\text{H}^+\text{H}^-}) \tag{1.51}$$

where the best values for the coefficients C_{ion} and C_{cov} are obtained by solving the relevant secular equation. This ground-state wave function improves the calculated energy of the H_2 molecule by 0.25 eV relative to the energy obtained from the covalent wave function.

The incorporation of ionic terms in molecular wave functions plays a major role in the description of solvent (and protein) effects on chemical processes. This point will be emphasized repeatedly throughout this book.

1.3.3. Semi-empirical Parametrization of the VB Hamiltonian

The VB potential surfaces ^1E and ^3E can be approximated by analytical Morse and anti-Morse functions:

$$^1E_+ \approx D[\exp\{-2a(R - R^0)\} - 2\exp\{-a(R - R^0)\}] = M(R)$$

$$^3E_- \approx \frac{1}{2}\,D[\exp\{-2a(R - R^0)\} + 2\exp\{-a(R - R^0)\}] = M^*(R) \tag{1.52}$$

The parameters D and a can be obtained by fitting ^1E to the actual ground state of the given molecule (D is determined by the observed bond dissociation energy and a is determined by the vibrational force constant). This allows one to express J and K in terms of available experimental information. That is, from eqs. (1.47), (1.49), and (1.52) we obtain

$$\bar{J} = [(M + M^*) + S^2(M - M^*)]/2$$

$$\bar{K} = [(M - M^*) + S^2(M + M^*)]/2 \tag{1.53}$$

where the bar indicates an effective integral. This powerful procedure is another example of the philosophy of semi-empirical parameterization which was introduced in Section 1.2.2. Here we evaluate the relevant integrals (e.g., J and K) using readily available experimental information [e.g., the observed bond properties for the function $M(R)$], rather than quantum mechanical wave functions. In general, if the approximate quantum mechanical treatment captures the correct physics of a given system, the corresponding semiempirical procedure is expected to provide an effective way of describing that system. This point will be emphasized in later chapters.

The covalent nature of the chemical bond changes significantly when the bonding electrons are shared between atoms of different electronegativity. In such cases the VB wave function should include ionic contributions of the type discussed in the previous section. For example, in describing the HF molecule we should consider the contribution from the covalent and ionic wave functions:

$$\psi_1 = \psi^{cov} = (H - F) = |\overline{\chi_F}\chi_H| + |\chi_F\overline{\chi_H}|$$

$$\psi_2 = \psi^{ion} = (H^+F^-) = |\chi_F\overline{\chi_F}| \qquad (1.54)$$

where χ_F is the unoccupied atomic orbital of the F atom. For simplicity we consider all the occupied orbitals of the F atom to form an inner effective core for the unoccupied orbital. With this basis set, we can write the total wave function as

$$\Psi = C_1\psi_1 + C_2\psi_2 \qquad (1.55)$$

and obtain the ground state energy, E_g, from the lowest eigenvalue of the secular equation,

$$\begin{vmatrix} H_{11} - E & H_{12} \\ H_{12} & H_{22} - E \end{vmatrix} = 0 \qquad (1.56)$$

where we have neglected the overlap integrals S_{12} and assume that their effect on the problem can be absorbed into the parameterization of H_{12}.

The potential surfaces E_g, H_{11}, and H_{22} of the HF molecule are described in Fig. 1.6. These potential surfaces provide an instructive example for further considerations of our semiempirical strategy (Ref. 5). That is, we would like to exploit the fact that H_{11} and H_{22} represent the energies of electronic configurations that have clear physical meanings (which can be easily described by empirical functions), to obtain an analytical expression for the off-diagonal matrix element H_{12}. To accomplish this task we represent H_{11}, H_{22}, and E_g by the analytical functions

$$H_{11} = \varepsilon_1 = \bar{M}(\bar{D}, R^0 + \delta, a)$$

$$= \bar{D}(\exp\{-2a(R - R^0 - \delta)\} - 2\exp\{-a(R - R^0 - \delta)\})$$

$$H_{22} = \varepsilon_2 = I - EA - 332/R + A\exp(-bR) + CR^{-9}$$

$$E_g = M(D, R^0, a) = D(\exp\{-2a(R - R^0)\} - 2\exp\{-a(R - R^0)\}) \qquad (1.57)$$

where now the energies and distances are given, respectively, in kcal/mol and Å. The potential surfaces of ε_1 and E_g are described by the same Morse potential function used in eq. (1.52), but with different parameters in each case (e.g., the R^0 for $\bar{M}$ is taken as the R^0 of M plus the increment δ). The

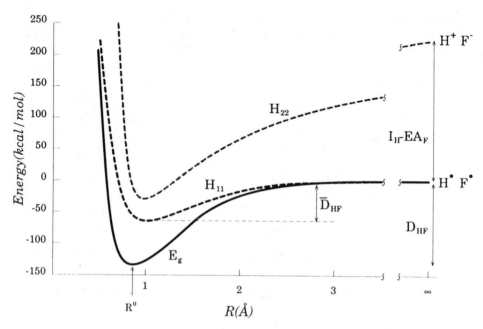

FIGURE 1.6. The VB potential surface for the HF molecule.

ionic potential surface ε_2 is described by a function that represents the sum of the Coulombic attraction between the ions and their short-range hard-core repulsion. The term (I-EA) is the energy of forming H^+ and F^- from $H^{\cdot}$ and $F^{\cdot}$ at infinite separation, where I and EA are, respectively, the ionization potential of H and the electron affinity of F. The parameters in eq. (1.57) can be obtained from experimental information by the following procedure. The parameters D and R_0 of M are obtained by using the observed bond energy and bond length for D and R_0, respectively. The parameter a is calibrated by adjusting its value to reproduce the observed vibrational frequency of the given bond [with the vibrational force constant from the second derivative of $M(R)$ one obtains $a = 0.12177\omega(\mu/D)$], where ω is the vibrational frequency of the given bond and μ is the corresponding reduced mass. (In this expression a is given in Å^{-1}, μ in au, D in cm^{-1}, and ω in cm^{-1}.) The parameter $\bar{D}$ in $\bar{M}$ is obtained through the relationship

$$\bar{D}_{XY} = (D_{XX}D_{YY})^{1/2} \tag{1.58}$$

where the D_{XX} is the dissociation energy of the indicated homonuclear diatomic molecule (e.g., H_2). The parameter δ is usually taken as 0.05 Å. The parameters A and C in the ionic term H_{22} can be estimated by requiring that the dipole moment obtained from the ground-state wave function reproduces the observed dipole moment (see exercise below) and that the minimum of ε_2 corresponds to the sum of the ionic radii of X and Y.

TABLE 1.1. Parameters for typical X–Y Bond Hamiltonians[a]

Morse Potential

X	Y	D	$\bar{D}$	R^0	a
H	H	104	104	0.74	1.99
O	O	54	54	1.32	2.32
N	N	58	58	1.40	2.30
C	C	88	88	1.54	1.60
F	F	36	36	1.42	2.98
Cl	Cl	58	58	1.99	2.03
Br	Br	47	47	2.28	1.94
I	I	36	36	2.67	1.85
S	S	51	51	2.08	1.83
H	O	102	75	0.96	2.26
H	C	106	96	1.09	1.80
H	N	103	78	1.00	2.07
H	F	134	61	0.92	2.27
H	Cl	102	78	1.28	1.90
H	Br	89	70	1.41	1.82
H	I	74	61	1.60	1.75
H	S	81	73	1.32	1.87
C	O	92	69	1.43	1.87
C	F	105	56	1.39	1.81
C	Cl	78	71	1.80	1.51
C	Br	66	64	1.94	1.58
C	I	54	56	2.14	1.56

Ionic State Potential

X^+	Y^-	A	b	C	I_X	EA_Y
H^+	$O^-[C_2H_5O^-]^b$	250	2.5	12	312	40
$C^+[CH_3^+]^b$	$O^-[CH_3O]^b$	5200	2.5	100	175	36
$C^+[C_2H_5^+]^b$	$O^-[C_2H_5O^-]^b$	5200	2.5	100	170	40
H^+	F^-	640	2.5	20	313	83
H^+	Cl^-	1000	2.5	25	313	87
H^+	Br^-	2000	2.5	30	313	82
H^+	I^-	1950	2.5	6400	313	76
H^+	$S^-[SH^-]^b$	100	2.5	975	313	48
H^+	$N^-[NH_2^-]^b$	10	2.5	50	313	6

[a]Energies in kcal/mol, distances in Å, all values are for a single bond (e.g., O–O not O=O).
[b]The corresponding molecular ion is in brackets.

With the approximated functions for ε_1 and ε_2 and with a Morse potential, M, that describes the observed properties of E_g we can solve eq. (1.56) and obtain

$$H_{12} = ((\varepsilon_1 - M)(\varepsilon_2 - M))^{1/2} \tag{1.59}$$

This gives a useful analytical approximation for H_{12}.

Table 1.1 gives the parameters for ε_1, ε_2, and E_g for representative bonds. With these parameters and eqs. (1.57–1.59) we can describe the bonding properties of many molecules, and more importantly (as will be demonstrated in the next chapter), we can consider bond-breaking reactions in solutions.

1.3.4. Molecular Dipole Moment

The charge distribution of neutral polar molecules is characterized by a dipole moment which is defined classically by $\boldsymbol{\mu} = \Sigma \, q_i \mathbf{r}_i$, where the molecular charge distribution is defined in terms of the residual charges (q_i) at the position $\mathbf{r}_i$. The observed molecular dipole moment provides useful information about the charge distribution of the ground state and its ionic character.

In order to compare calculated and observed dipole moments, we should replace the classical expression of the dipole moment by its quantum analogue $\boldsymbol{\mu} = \int \Psi^* \bar{\boldsymbol{\mu}} \Psi \, d\tau$ where $\bar{\mu}$ is the dipole moment operator (given by $\bar{\boldsymbol{\mu}} = -\Sigma_i \, e\mathbf{r}_i + \Sigma_j \, eZ_j\mathbf{R}_j$ with i and j running over the electronic and nuclear coordinates, respectively, and $-e$ the electron charge). The actual calculation of a VB dipole moment is described below.

Exercise 1.7. Evaluate the dipole moment for the HF molecule using the parameter of Table 1.1, and examine the change of the calculated dipole moment with the change of A (e.g., $A = 640$ and 2000).

Solution 1.7. Using the given H_{11}, H_{22}, and H_{12} at $R = R^0 = 0.92$ Å we construct the **H** matrix, send it to the diagonalization subroutine DIAG, and obtain the eigenvalues ε_1 and ε_2 [where ε_1 is the lowest eigenvalue (E_g)] and the corresponding eigenvectors C_1 and C_2. The ground-state dipole moment can now be evaluated by

$$
\begin{aligned}
\boldsymbol{\mu}_g &= \int \Psi_g^* \bar{\boldsymbol{\mu}} \Psi_g \, d\tau \\
&= C_{11}^2 \bar{\boldsymbol{\mu}}_{11} + C_{12}^2 \bar{\boldsymbol{\mu}}_{22} + 2 C_{11} C_{12} \bar{\boldsymbol{\mu}}_{12} \\
&\approx 0 + C_{12}^2 \boldsymbol{\mu}(H^+ F^-) + 0 \\
&= C_{12}^2 e \mathbf{R}_{HF}^0
\end{aligned} \tag{1.60}
$$

where $\bar{\boldsymbol{\mu}}$ is the dipole operator and $\bar{\boldsymbol{\mu}}_{ij} = \int \psi_i \bar{\boldsymbol{\mu}} \psi_j \, d\tau$. Here we use the fact that ψ_1 is a covalent state with a zero dipole moment and ψ_2 is a fully ionic state with plus and minus electron charge (e) on the centers of H and F respectively. The integral $\bar{\mu}_{12}$, which is related to the overlap between ψ_1 and ψ_2, is assumed to be negligible. The dipole moment is usually given in units of Debye (D), where the dipole formed by a positive and a negative charge, of 1 au each, which are separated by 1 Å, has the value of 4.803 D. The observed dipole moment of H–F is 1.82 D. This value should be obtained using the parameters of Table 1.1.

It is instructive to consider the effect of an external charge on the HF molecule. This is conveniently done by evaluating the interaction between the ion pairs and an external charge at site C.

$$\varepsilon_2' = \varepsilon_2 + 1/R_{HC} - 1/R_{FC} \approx \varepsilon_2 - \mu_2 \xi_C \tag{1.61}$$

where μ_2 corresponds to the μ_{22} of eq. (1.60) and ξ_C is the field from the charge on this dipole. We note that the external charge at R_C does not change the energy of the covalent ε_1 state and we further assume that H_{12} is unperturbed by the external charge (this assumption is justified if we neglect the overlap between orbitals on C and the orbitals of our molecule). Thus, all that is required to estimate the interaction of the HF molecule with an external charge is to rediagonalize eq. (1.56) with the new ε_2. In the case where $|\varepsilon_2 - \varepsilon_1| \gg |H_{12}|$ and $|\varepsilon_2 - \varepsilon_1| \gg |\mu_2 \xi_C|$ we obtain

$$\begin{aligned} E_g' &= \frac{1}{2} [(\varepsilon_2' + \varepsilon_1) - ((\varepsilon_2' - \varepsilon_1)^2 + 4H_{12}^2)^{1/2}] \\ &\simeq \varepsilon_1 - (H_{12}^2/(\varepsilon_2' - \varepsilon_1)) \simeq \varepsilon_1 - (H_{12}^2/(\varepsilon_2 - \varepsilon_1))(1 - (\varepsilon_2' - \varepsilon_2)/(\varepsilon_2 - \varepsilon_1)) \\ &\simeq E_g - (H_{12}/(\varepsilon_2 - \varepsilon_1))^2 \mu_2 \xi_C = E_g - C_{12}^2 \mu_2 \xi_C = E_g - \mu_g \xi_C \end{aligned} \tag{1.62}$$

where we use simple perturbation expansions ($\sqrt{1+x} \simeq 1 + x/2$) and $1/(A + x) \simeq A^{-1}(1 - x/A)$ and use the fact that C_{12} can be approximated by $H_{12}/(\varepsilon_2 - \varepsilon_1)$ when $\varepsilon_2 - \varepsilon_1 \gg |H_{12}|$. This derivation demonstrates that the incorporation of the interaction between ψ_2 and the external charge in ε_2 is approximately equivalent to the consideration of the interaction between the ground-state dipole moment and the external charge.

Exercise 1.8. Plot E_g as a function of R for the HF molecule in the presence of an external charge 10 Å from the F atom.

Solution 1.8. See Program 1.C.

1.4. SMALL POLYATOMIC MOLECULES

The approach described above for diatomic molecules can be extended to polyatomic molecules. We will outline here VB treatments and consider MO approaches only in a few selected cases in subsequent chapters.

While rigorous VB treatment for polyatomic molecules is quite complicated, one can exploit simplifying approximations quite effectively. In particular, it is simple to describe the energies of various VB wave functions by the *"perfect-pairing"* approximation which is a generalization of eqs. (1.47) and (1.49). This approximation describes the energies of a given bonding arrangement or *resonance structure* (ψ_n) by

$$H_{nn} = \varepsilon_n = \sum_i A_i^0 + \sum_{ij} J_{ij} + \sum_{ij} C_{ij}^n K_{ij} \tag{1.63}$$

where A_n^0 is the atomic energy of the ith atom, while J_{ij} and K_{ij} are the Coulomb and exchange integrals for the interaction between the i and j pair of atoms. The coefficients C_{ij}^n that scale the exchange integrals reflect the correlation between the corresponding spins. For antiparallel and parallel spin, the C's are $+1$ and -1 respectively. This accounts for the corresponding $J + K$ and $J - K$ bond energy contributions in eqs. (1.47) and (1.49). When the spins are uncorrelated, there are three possible spin wave functions with parallel spins and only one with antiparallel spins. The corresponding exchange contribution ($\frac{1}{4}[K + 3(-K)] = -\frac{1}{2}K$) is reproduced by $C = -\frac{1}{2}$. The J's and K's in eq. (1.63) can be determined semiempirically from eq. (1.53). The off-diagonal elements for interaction between different resonance structures are determined by more complicated diagramatic approaches which are outlined elsewhere (Ref. 4). Here we will simply use the results obtained by this approach for our specific demonstration.

As a demonstration of the VB approach, consider the reaction

$$CH_4 + Cl \rightarrow CH_3 + HCl \tag{1.64}$$

This reaction can be described by the two VB resonance structures

$$\psi_1 = [C - H \quad Cl \quad P]$$

$$\psi_2 = [C \quad H - Cl \quad P] \tag{1.65}$$

where we added here a fourth "phantom atom" P so that the problem can be treated first as a formal four-electron problem, and then converted to a three-electron problem by removing P to infinity and setting to zero all the integrals that involve P. The atom C designates here a carbon atom with three attached hydrogens. The corresponding diagonal energies are

$$H_{11} = \varepsilon_1 = J + K_{12} + K_{34} - \frac{1}{2} K_{23} - \frac{1}{2} K_{13} - \frac{1}{2} K_{14} - \frac{1}{2} K_{24}$$

$$H_{22} = \varepsilon_2 = J + K_{23} + K_{14} - \frac{1}{2} K_{12} - \frac{1}{2} K_{13} - \frac{1}{2} K_{24} - \frac{1}{2} K_{34} \tag{1.66}$$

where $J = J_{12} + J_{13} + J_{14} + J_{23} + J_{24} + J_{34}$. The indices 1–4 refer to the four atoms in the order of eq. (1.65) while the K's and J's are given in eq. (1.53). Using the diagramatic approach described elsewhere (e.g., Ref. 4), we can also obtain

$$H_{12} = -\frac{1}{2} (J + K_{12} + K_{23} + K_{34} + K_{14} - 2K_{13} - 2K_{24})$$

$$S_{12} = 1/2 \tag{1.67}$$

Diagonalizing the corresponding 2×2 secular equation and some algebraic manipulation gives the four-electron ground-state potential surface

$$E_g = J - \left(\frac{1}{2} \{ (\alpha - \beta)^2 + (\beta - \gamma)^2 + (\gamma - \alpha)^2 \} \right)^{1/2} \tag{1.68}$$

where $\alpha = K_{12} + K_{34}$, $\beta = K_{14} + K_{23}$, and $\gamma = K_{13} + K_{24}$. Taking the phantom atom to infinity gives the potential surface for our reaction

$$E_g = \left[J_{12} + J_{23} + J_{13} \right.$$
$$\left. - \left\{ \frac{1}{2} [(K_{12} - K_{23})^2 + (K_{23} - K_{13})^2 + (K_{13} - K_{12})^2] \right\}^{1/2} \right] \frac{1}{1 + \bar{S}^2}$$
$$\tag{1.69}$$

where the $\bar{S}$ term [which represents the effective overlap of eq. (1.48)] is not obtained from the treatment described above but is needed for a consistent semi-empirical treatment (see, for example, Ref. 6). The value of $\bar{S} = 0.424$ provides a good approximation for the potential surface at the transition state region.

Equation (1.69) can be expressed analytically using the semi-empirical approximation of eq. (1.53) for J and K. The resulting surface for the colinear reaction with $r_{12} + r_{23} = r_{13}$ (where 1, 2, and 3 describe our three reacting atoms) is shown in Fig. 1.7. The main features of this surface can be discussed in terms of the regions in the figure. At the reactant region r_{23} is very large and the surface is given approximately by $E'_g \approx \varepsilon_1 \approx M(r_{12})$ where E'_g is the ground-state energy relative to the reference energy of the $\dot{C}H_3 + H + Cl$ fragments. Similarly in the product region, the surface can be approximated by the Morse potential of the HCl molecule. In region b the nonbonded interaction between the C–H fragment and the Cl atom start to contribute due to the $M^*(r_{23})$ term of ε_1, and finally at region c, which is called the *transition state* region, $\varepsilon_1 \approx \varepsilon_2$ and the potential surface is given approximately by $E'_g \approx \frac{1}{2}(\varepsilon_1 + \varepsilon_2) - H_{12}$. The behavior of our semi-empirical surface at the transition state region is only an approximation of the true surface. However, this semiempirical surface gives a very reliable approxi-

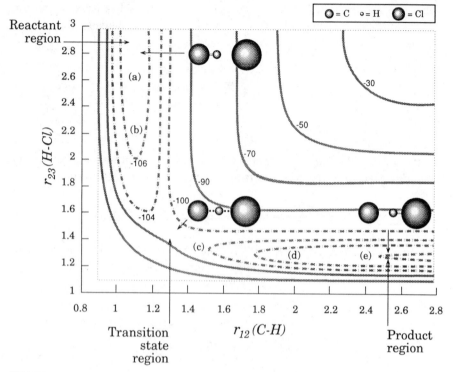

FIGURE 1.7. The potential energy surface of the $CH_4 + Cl$ supersystem for the collinear hydrogen abstraction reaction: $CH_4 + Cl \rightarrow CH_3 + HCl$. The counter lines are given in spaces of 10 kcal/mol and the coordinates in angstroms.

mation for the exact Born–Oppenheimer surface at the asymptotic regions (a) and (e). With this in mind, we can improve the surface by treating H_{12} as an adjustable function that is fitted to the available experimental information about the transition state region.

Exercise 1.9. Evaluate the potential surface for the $H + H_2 \rightarrow H_2 + H$ exchange reaction and determine the energy of the *transition state* obtained with $r_{12} = r_{23} = 1.4 \text{ Å}$ relative to the minimum energy of the system when one hydrogen atom is at infinity.

Solution 1.9. Use Program 1.D.

The same type of approach presented here can be used to describe chemical reactions of much larger molecules as long as the reaction region is restricted to only a few atoms. This point will be emphasized in subsequent chapters.

1.5. APPENDIX A—MOLECULAR ORBITAL TREATMENT OF MANY-ELECTRON SYSTEMS

The brief discussion of molecular orbitals given in this chapter is not sufficient background for actually applying the molecular orbital method in studies of molecular forces. This appendix supplies additional advanced material which will allow the reader to compare our VB examples to reasonable MO treatments.

The MO wave function for $2n$ electrons in n molecular orbitals is given by the following Slater determinant:

$$\Psi^{MO} = |\psi_1 \bar{\psi}_1 \psi_2 \bar{\psi}_2 \cdots \psi_n \bar{\psi}_n| \qquad (1.70)$$

where the ψ_1 are expressed as a sum of n atomic orbitals $\psi_1 = \Sigma_i \, C_{i\mu} \chi_\mu$. The Hamiltonian in the many-electron case involves the electron–electron repulsion term and is given by

$$\mathbf{H} = -\sum_i \nabla_i^2/2 - \sum_i \sum_A Z_A/r_{iA} + \sum_{i<j} 1/r_{ij} = \sum_i h(i) + \sum_{i<j} 1/r_{ij} \quad (1.71)$$

where i runs over the $2n$ electrons of the system and A runs over the nuclear cores. The exact treatment of the many-electron problem is quite involved and is not essential for following the arguments of this book (see, for example, Ref. 2 for more complete derivations). Thus, we only give a summary of results which will be useful for various considerations presented in later stages.

The total energy associated with the Slater determinant of eq. (1.70) is given by (see Ref. 2)

$$E = 2 \sum_i^n \varepsilon_i^0 + \sum_{i,j}^n (2J_{ij} - K_{ij}) \qquad (1.72)$$

where ε_i^0 is the energy of the molecular orbitals in the zero-order single particle Hamiltonian $h(i)$ which does not include the $(1/r_{ij})$ term

$$\varepsilon_i^0 = \int \psi_i^* h(i) \psi_i \, d\tau_i = \langle i|h|i \rangle \qquad (1.73)$$

where we introduce the useful Dirac ($\langle \; | \; \rangle$) notation for our integral.

The quantities J and K are called, respectively, the *Coulomb* and *exchange* integrals and are given by

$$J_{jk} = \int \psi_j^*(1)\psi_j(1)r_{12}^{-1}\psi_k^*(2)\psi_k(2) \, d\tau_1 \, d\tau_2$$

$$K_{jk} = \int \psi_j^*(1)\psi_k(1)r_{12}^{-1}\psi_j(2)\psi_k^*(2) \, d\tau_1 \, d\tau_2 \qquad (1.74)$$

Minimizing the total energy E with respect to the MO coefficients (see Refs. 2 and 3) leads to the matrix equation $\mathbf{FC} = \mathbf{SCE}$ (where $\mathbf{S}$ is the overlap matrix). Solving this matrix is called the *self-consistent field* (SCF) treatment. This is considered here only on a very approximate level as a guide for qualitative treatments (leaving the more quantitative considerations to the VB method). The SCF–MO derivation in the *zero-differential overlap* approximations, where overlap between orbitals on different atoms is neglected, leads to the secular equation

$$\mathbf{FC}_i = \varepsilon_i \mathbf{C}_i \tag{1.75}$$

where the relevant matrix elements are given by

$$F_{\mu\mu} = H_{\mu\mu}^{\text{core}} - \frac{1}{2} P_{\mu\mu}\gamma_{AA} + \sum_B P_{BB}\gamma_{AB}$$

$$F_{\mu\nu} = H_{\mu\nu}^{\text{core}} - \frac{1}{2} P_{\mu\nu}\gamma_{\mu\nu} = \beta_{\mu\nu} - \frac{1}{2} P_{\mu\nu}\gamma_{\mu\nu}$$

$$\gamma_{\mu\nu} = \int |\chi_\mu(1)|^2 r_{12}^{-1} |\chi_\nu(2)|^2 \, d\tau_1 \, d\tau_2 \tag{1.76}$$

here the χ's are the indicated atomic orbitals, H^{core} is the operator h in eq. (1.71) and

$$H_{\mu\mu}^{\text{core}} = \langle \mu | h | \mu \rangle \tag{1.77}$$

γ is called the *electron–electron repulsion integral* and P is referred to as the bond order and is given by

$$P_{\mu\nu} = \sum_i n_i C_\mu^i C_\nu^i$$

$$P_{BB} = \sum_\mu^B P_{\mu\mu} \tag{1.78}$$

where n_i is the number of electrons in the ith orbital and P_{BB} is the total bond order on atom B. Equation (1.75) is an iterative SCF equation since $F_{\mu\nu}$ depends on $P_{\mu\nu}$, which in turn depends on the $\mathbf{C}$ vector.

The integral $H_{\mu\mu}^{\text{core}}$ can be written as

$$H_{\mu\mu}^{\text{core}} = U_{\mu\mu} - \sum_B \langle \mu | V_B | \mu \rangle = U_{\mu\mu} - \sum_B V_{AB} \tag{1.79}$$

where $-V_B$ is the potential due to the nucleus and inner shells of atom B, while $U_{\mu\mu}$ is the one center term,

$$U_{\mu\mu} = \langle \mu | -\frac{1}{2} \nabla^2 - V_A | \mu \rangle \tag{1.80}$$

and is essentially an atomic quantity. Approximating V_{AB} by $Z_B \gamma_{AB}$ gives

$$F_{\mu\mu} = U_{\mu\mu} + \left[P_{AA} - \frac{1}{2} P_{\mu\mu} \right] \gamma_{AA} - \sum_B Q_B \gamma_{AB} \qquad (1.81)$$

where Q_B is the net charge on atom B given by $Q_B = (Z_B - P_{BB})$. The potential surface of eq. (1.72) is now given by

$$E = \frac{1}{2} \sum_{\mu v} P_{\mu v}(H_{\mu v}^{\text{core}} + F_{\mu v}) + \sum_{A>B} Z_A Z_B R_{AB}^{-1} \qquad (1.82)$$

where the first term is the electronic energy of eq. (1.72) and the last term is the nuclear interaction term. The potential E can be also written as

$$E = \sum_A E_A + \sum_{A<B} E_{AB} \qquad (1.83)$$

where E_A and E_{AB} are, respectively, the one- and two-body interaction terms, given by

$$E_A = \sum_\mu^A P_{\mu\mu} U_{\mu\mu} + \frac{1}{2} \sum_{\mu v}^A \left(P_{\mu\mu} P_{vv} - \frac{1}{2} P_{\mu v}^2 \right) \gamma_{AA}$$

$$E_{AB} \approx \sum_\mu^A \sum_v^B \left(2P_{\mu v} \beta_{\mu v} - \frac{1}{2} P_{\mu v}^2 \gamma_{AB} \right) + \sum_{A<B} Q_A Q_B R_{AB}^{-1} \qquad (1.84)$$

where we use the approximation $\gamma_{AB} = R_{AB}^{-1}$ (which is a good approximation for large values of R_{AB}).

To simplify the discussion, we consider only the first two iterations of eq. (1.75), starting with $P_{\mu v}^{(0)} = \delta_{\mu v}$, and then solving once to obtain the coefficients **C**. We then use them with eq. (1.78) to construct $P_{\mu v}^{(1)}$ and to evaluate eq. (1.81), which is then resubstituted in eq. (1.75) to give $P_{\mu v}^{(2)}$, which is then used to evaluate the energy of eq. (1.84).

To simplify the exercises in the book, we will write the energy in the form

$$E = \sum_{\mu v} P_{\mu v} F_{\mu v} + \sum_{A<B} Z_A Z_B R_{AB}^{-1} + E' = \sum_i n_i \varepsilon_i + \sum_{A<B} Z_A Z_B R_{AB}^{-1} + E' \qquad (1.85)$$

where E' includes all of the remaining terms. Examining $E - E'$ rather than the total energy E will give us a qualitative description of the corresponding potential surfaces in terms of the orbital energies ε which are directly obtained from eq. (1.75).

As an example of the MO treatment, let us consider the case of a diatomic molecule (AB) with two electrons in the presence of a bare proton. For our specific example, we will neglect the resonance interaction between the molecule and the ion using $\beta_{AC} = \beta_{BC} = 0$, where C designates an atomic orbital on the proton site.

The ε_i for our two-electron system are obtained now by

$$\begin{vmatrix} F_{AA} - \varepsilon & \beta_{AB} \\ \beta_{AB} & F_{BB} - \varepsilon \end{vmatrix} = \begin{vmatrix} \alpha_A - Q_C\gamma_{AC} - \varepsilon & \beta_{AB} \\ \beta_{AB} & \alpha_B - Q_C\gamma_{BC} - \varepsilon \end{vmatrix} = 0 \qquad (1.86)$$

where $\alpha_A = U_{\mu\mu} + \gamma_{AA}/2$. Here we use the fact that the first iteration of the present approximation gives $Q_A^0 = Q_B^0 = 0$, since $P^0 = 1$ and $Z = 1$ for the A and B atoms. The interesting new feature in our model is the fact that the effective ionization potential α_μ is modified by the potential from the external charge (see Fig. 1.3 for a related problem). This leads to further polarization of the charges $Q_A^{(1)}$ and $Q_B^{(1)}$ [obtained by solving eq. (1.75) and using the solution to evaluate $P_{AA}^{(1)}$ and $P_{BB}^{(1)}$] relative to their value at the absence of Q_C. The corresponding lowest ε is given by the standard solution of a 2×2 secular equation:

$$\varepsilon_1 = (1/2)[(F_{AA} + F_{BB}) - ((F_{AA} - F_{BB})^2 + 4\beta_{AB}^2)^{1/2}] \qquad (1.87)$$

and for $\alpha_A > \alpha_B$ we obtain from (1.85) a ground state of the type described in Fig. 1.4.

The interaction between the proton charge Q_C and the molecule follows a simple physical trend; the positively charged ion polarizes the molecule by changing the ionization energies and then interacting with the resulting polar molecule through the $(Q_A Q_C R_{AC}^{-1} + Q_B Q_C R_{BC}^{-1})$ term of eq. (1.84).

Exercise 1.10. Evaluate the charge Q_{AA} for $\alpha_A = -13.6$ eV, $\alpha_B = -15.5$ eV, $\beta_{AB} = -6$ eV, $\gamma_{AC} = 12$ eV by solving the equation $\mathbf{FC} = \varepsilon\mathbf{C}$ with the $\mathbf{F}$ of eq. (1.86). Use the ε_1 of eq. (1.87) to obtain the $\mathbf{C}$'s vector and then use eq. (1.78) to evaluate P_{AA} (placing two electrons in orbital 1, i.e. $n_1 = 2$).

Exercise 1.11. Evaluate a simplified potential surface of the form shown in Fig. 1.4 using the term $E - E'$ of eq. (1.85). To do this, consider our diatomic molecule in the presence of a proton using $\alpha_A = -13.6$ eV $\alpha_B = -15.5$ eV, $\beta_{AB} = -6 \exp[-2R]$ eV (R in Å) $\gamma_{AB} = 1/R$. Solve eq. (1.86) for $R = 1.5$, 2.0, and 3.0 and plot the $E - E'$ of eq. (1.85).

Solution 1.11. Use Program 1.B with the SCF rather than the Huckel option.

The MO approach described above can give useful qualitative insight. However, it encounters major problems in describing the electron pairs that form chemical bonds. For example, when we break the H_2 molecules to two radicals, we obtain the SCF energy of $(H \cdot\cdot H)$. This energy is very different than the correct energy of two isolated hydrogen atoms $(2\alpha_H)$ since the correlated motion of the electrons is not described correctly. The electron-correlation problem can be corrected by configuration interaction (CI) treatment, which is not considered here, but this is done at the expense of simplicity and physical insight.

1.6.   SOME RELEVANT COMPUTER PROGRAMS

1.A.   An Example of Using a Diagonalization Subroutine

```
c          Here we simply solve a 2x2 matrix so that the solution can be
c          verified by the reader. But subroutine DIAG can be used for
c          much larger matrices.

           implicit real*8 (a-h,o-z)
           dimension h(20,20),c(20,20),ev(20)
           n=2
           h(1,1)=10.
           h(2,2)=20.
           h(1,2)=2.
           h(2,1)=2.
           call diag(h,c,ev,n)
           write(6,*) 'eigenvalues  eigenvectors'
           do i=1,n
                print 10,ev(i),(c(i,j),j=1,n)
           enddo
10         format(2x,f10.4,7x,6f10.4)
           end
c=========================================================
           subroutine diag(h,c,ev,n)

c          Using the cyclic Jocobi's method for eigenvalues of symmetric matrix.
c          See G. Dahlquist and A. Bjorck, Numerical Methods, Prentice Hall, Inc. 1974.

c          h          The matrix to be diagonalized
c          n          The order of the h matrix
c          c          The eigenvectors of the matrix
c          ev         The eigenvalues of the h matrix
           implicit real*8 (a-h,o-z)
           dimension h(20,20),c(20,20),ev(20)
           dimension ht(20,20),ct(20,20)
           l=n-1
           do 1 i=1,l                                ! Initializes of the c matrix as a unity matrix.
                ia=i+1
                c(i,i)=1.0
                do 1 j=ia,n
                     c(j,i)=0.0
1          c(i,j)=c(j,i)
           c(n,n)=1.0
2          do 3 i=1,l
                ia=i+1
                do 3 j=ia,n
                     if(dabs(h(i,j)).ge.1.d-8) then
                          d=-h(i,j)
                          uu=(h(i,i)-h(j,j))*.5
                          w=dsqrt(d*d+uu*uu)
                          b=d/dabs(d)
                          ad=dabs(5.e-1-(b*uu)/(2*w))
                          si=dsqrt(ad)
```

```
                    ae=dabs(1.0-ad)
                    co=dsqrt(ae)
                    d=h(i,j)
                    uu=h(i,i)
                    w=h(j,j)
                    do 4 k=1,n                          ! Updates the ith column and the jth row
                         if(k.ne.i.and.k.ne.j) then     ! of the H matrix.
                              aa=h(k,i)
                              ab=h(k,j)
                              sa=c(k,i)
                              sb=c(k,j)
                              h(k,i)=aa*co-ab*si
                              h(i,k)=h(k,i)
                              h(k,j)=aa*si+ab*co
                              h(j,k)=h(k,j)
                              c(k,i)=sa*co-sb*si
                              c(k,j)=sa*si+sb*co
                         endif
4                   continue
                    ac=si*co
                    h(i,i)=uu*ae+w*ad-2*d*ac
                    h(j,j)=uu*ad+w*ae+2*d*ac             ! Evaluates the eigenvalues.
                    h(j,i)=0.0
                    h(i,j)=h(j,i)
                    sa=c(i,i)
                    sb=c(i,j)
                    c(i,i)=sa*co-sb*si                  ! Evaluates the eigenvectors
                    c(i,j)=sa*si+sb*co
                    sa=c(j,i)
                    sb=c(j,j)
                    c(j,i)=sa*co-sb*si
                    c(j,j)=sa*si+sb*co
               endif
3        continue
         a=0.0
         do 5 i=1,l                                     ! Checks the convergence of
              ia=i+1                                    ! the cyclic Jacobi's method.
              do 5 j=ia,n
5        a=a+dabs(h(i,j))
         ab=.5*n*(n-1)
         a=a/ab
         if(a.gt.1.d-8) go to 2
         do 30 j=1,n                                    ! Arranges the eigenvalue
              emx=-10000.                               ! according to their magnitude
              do 31 i=1,n                               ! ( the largest first ).
                   et=h(i,i)
                   if(et.gt.emx) then
                        lt=i
                        emx=et
                   endif
31            continue
              ht(j,j)=emx
              h(ilt,ilt)=-100000.
              do 33 k=1,n
```

```
33          ct(k,j)=c(k,ilt)
30          continue
            do 40 i=1,n
                it=i
                ev(it)=ht(i,i)
                do 40 k=1,n
40          c(k,it)=ct(k,i)
            return
            end
```

1.B. Molecular Orbitals Calculations by the Huckel (or the SCF) Method

```
c           This program calculates  potential surfaces by the Huckel or the SCF approximations.
c           The calculations are done with one orbital per atom.

c       h           The Huckel matrix ( or the SCF matrix )
c       natom       The number of atoms
c       n           The order of the h matrix
c       p           The bond order matrix ( p0 and p1 are used for updating this matrix )
c       x           The coordinate vector
c       c           The eigenvectors of the h matrix
c       z           The core charge
c       nn          The number of electrons in each MO
c       code        The mode of calculations ('SCF' or 'huckel')
        implicit real*8 (a-h,o-z)
        character code*8
        dimension h(20,20),c(20,20),p(20,20),r(20,20),q(20),ev(20)
        dimension p1(20,20),p0(20,20),hs(20,20)
        dimension nn(20),et(100),x(30,3)
        common /parm/alf0(20),z(20),beta,d,fmu,p
        common /ind/xplol(3000,3),fpol(3000,3),vq(30),x0(3),npol
        read (5,*) code                                 ! See data at the end of this program.
        read (5,*) natom
        n=natom
        read (5,*) a
        read (5,*) cc
        read (5,*) aa
        read (5,*) ((x(i,k),k=1,3),i=1,natom)
        read (5,*) (alf0(k),k=1,n)
        read (5,*) (z(k),k=1,n)
        read (5,*) ((p1(i,k),k=1,n),i=1,n)
        read (5,*) beta,d,fmu
        read (5,*) (nn(k),k=1,n)
        read (5,*) isol
        do i=1,n                                        ! p0(i,j) is the initial value of p1(i,j).
            do j=1,n
                p0(i,j)=p1(i,j)
            enddo
        enddo
```

```
c       Calculates the energy as a function of the interatomic distance.
        do 1000 ii=1,20
            do i=1,n                                          ! Sets p1(i,j) to its initial value after
                do j=1,n                                      ! each iteration.
                    p1(i,j)=p0(i,j)
                enddo
            enddo
            x(1,1)=x(1,1)-0.3                                 ! Changes the bond length.
            do i=1,natom                                      ! Calculates interatomic distances.
                do j=1,natom
                    r2=0.
                    do k=1,3
                        r2=r2+(x(i,k)-x(j,k))**2
                    enddo
                    r(i,j)=sqrt(r2)
                enddo
            enddo
            nloop=12
            emin=1000.
            do i=1,natom                                      ! Sets the initial charges to zero.
                q(i)=0.
            enddo
            do l=1,nloop
                if ( code.eq.'huckel') then
                    call huckel_mat(h,r,q,n,a)
                else
                    call scf_mat(h,r,q,n,a)
                endif
                if(l.eq.nloop) then
                    print 101
                    do i=1,n
                        print 102,(h(i,j),j=1,n)
                    enddo
                endif
                do i=1,n
                    do j=1,n
                        hs(i,j)=h(i,j)
                    enddo
                enddo
                call diag(h,c,ev,n)
                call pmat(c,p1,q,nn,n,aa)
                if(l.eq.nloop) then
                    print 104
                    do j=1,n
                        print 102,(c(i,j),i=1,n)
                    enddo
                    print 100
                    do j=1,n
                        print 102, (p(i,j),i=1,n)
                    enddo
                endif
                if(code.eq.'huckel') then
```

```
                  etot=0.0
                  do i=1,n                                    ! Calculates the total energy by
                      etot=etot+nn(i)*ev(i)                   ! the Huckel approximation.
                      do j=i+1,n
                          de=14.4*z(i)*z(j)/r(i,j)
    &                        -14.4*p(i,i)*p(j,j)/(d+r(i,j))
                          etot=etot+de
                      enddo
                  enddo
              else
                  etot=0.0
                  do i=1,n                                    ! Calculates the total energy by
                      etot=etot+p(i,i)*alf0(i)                ! the SCF approximation.
                      do j=i+1,n
                          de=14.4*(p(i,i)*p(j,j)-0.5*p(i,j)**2)/(2.0*d)
    &                        +2.*hs(i,j)+14.4*q(i)*q(j)/(r(i,j)+d)
    &                        +14.4*p(i,j)**2/(d+r(i,j))*0.5
    &                        +cc/(r(i,j)*exp(fmu*r(i,j)))
                          etot=etot+de
                      enddo
                  enddo
              endif
              if(etot.lt.emin) then
                      emin=etot
              endif
              et(l)=etot
          enddo
          write(6,117) x(2,1) - x(1,1)
          write(6,116) emin-alf0(1)-alf0(2)                   ! Shifts the energy to zero at infinity.
1000      continue
100       format(2x,'the p matrix')
101       format(2x,'the huckel matrix')
102       format(2x,3f10.3)
104       format(2x,'the MO eigenvectors')
117       format(2x,'R12',f12.6)
116       format(2x,'total energy',f12.6)
          end
c===========================================================
          subroutine huckel_mat(h,r,q,n,a)

c         Constructs the elements of the Huckel matrix.

          implicit real*8 (a-h,o-z)
          dimension h(20,20),r(20,20),q(20),p(20,20)
          common/parm/alf0(20),z(20),beta,d,fmu,p
          do 20 i=1,n
              do 10 j=1,n
                  if(i.eq.j) then
                      h(i,i)=alf0(i)-a*q(i)
                      do 5 k=1,n
                          if(i.ne.k) h(i,i)=h(i,i)-14.4*q(k)/(d+r(i,k))
5                         continue
                  else
```

```fortran
                        h(i,j)=beta*exp(-fmu*r(i,j))
                    endif
10              continue
20          continue
            return
            end
```

c===

```fortran
            subroutine scf_mat(h,r,q,n,a)
```

c   Constructs the elements of the SCF matrix.

```fortran
            implicit real*8 (a-h,o-z)
            dimension h(20,20),r(20,20),q(20),p(20,20)
            common/parm/alf0(20),z(20),beta,d,fmu,p
            do 20 i=1,n
                do 10 j=1,n
                    if(i.eq.j) then
                        h(i,i)=alf0(i)+14.4*p(i,i)/(2.*d)
                        do 5 k=1,n
                            if(i.ne.k) h(i,i)=h(i,i)-14.4*q(k)/(d+r(i,k))
5                       continue
                    else
                        h(i,j)=beta*exp(-fmu*r(i,j))-14.4*P(i,j)/(r(i,j)+d)*.5
                    endif
10              continue
20          continue
            return
            end
```

c===

```fortran
            subroutine pmat(c,p1,q,nn,n,aa)
```

c   Constructs the elements of the bond order matrix

```fortran
            implicit real*8 (a-h,o-z)
            dimension c(20,20),p(20,20),q(20),nn(20)
            dimension p1(20,20)
            common/parm/alf0(20),z(20),beta,d,fmu,p
            do 40 i=1,n
                do 40 j=1,n
                    p(i,j)=0.
                    do 30 k=1,n
                        p(i,j)=p(i,j)+nn(k)*c(i,k)*c(j,k)
30                  continue
40          continue
            do 51 i=1,n
                do 50 j=1,n                     ! Here we change P in a gradual way
                    p(i,j)=p1(i,j)+aa*(p(i,j)-p1(i,j))   ! to obtain a good SCF convergence.
                    p1(i,j)=p(i,j)
50              continue
                q(i)=z(i)-p(i,i)
51          continue
            return
            end
```

```
c================================================================
        subroutine DIAG
        use the subroutine from 1.A
c================================================================
c================================================================
```

Data for (1.B) the Huckel calculations of H_2 molecule

```
'huckel'
2     natom
4.0   a
0.    cc
0.01  aa
0.0   0.0   0.0     coordinates(x,y,z for each atom)
0.35  0.0   0.0
-13.6 -13.6     alf0
1.  1.     z
1.0 0.0     initial bond order matrix
0.0 1.0
-9.5 0.3 1.0 beta  d  mu
0 2   nn    electronic occupation
0     isol
```

Data for (1.B) MO calculations of H_2 in the presence of Li^+

```
'huckel'
3     natom
4.0   a
0.    cc
0.01  aa
0.00  0.0  0.0     coordinates
1.43  0.0  0.0
4.43  0.0  0.0
-13.6 -13.6 -6.    alf0
1.  1.  1.     z
1.0 0.0 0.0     initial bond order matrix
0.0 1.0 0.0
0.0 0.0 0.0
-9.5 0.3 1.0 beta  d  mu
0 0 2   nn    electronic occupation. Note that the last orbital is the one with the lowest energy.
0     isol
```

1.C.   A Simple VB Program

```
c        Evaluates potential surfaces for bond breaking reactions.

c        e1      The energy of the pure covalent state.
c        e2      The energy of the pure ionic state.
c        e       The total energy
c        dc      The dissociation energy of the purely covalent resonance form.
c        r0      The gas-phase equilibrium bond length for the x-y molecule.
c        r       The x-y bond length.
c        i       Ionization potential.
c        ea      Electron affinity.
```

```
c       d           The dissociation energy of the real bond.
c       delta       The shift in the minimum of the pure covalent state.
        real dc,r,r0,a,e1,e2,i,ea,d,a1,m,h12,e,delta
        print*,'input dc,r0,a,i,ea,c,d,a1'
        read*,dc,r0,a,i,ea,c,d,a1
        r=0.5
        delta=.1
        print10
        do l=1,50
            e1=dc*(exp(-2.*a*(r-r0-delta))-2.*exp(-a*(r-r0-delta)))
            e2=i-ea-332./r+a1*exp(-2.5*r)+c*r**(-9)
            m=d*(exp(-2.*a*(r-r0))-2.*exp(-a*(r-r0)))
            h12=(((e1-m)*(e2-m))**0.5
            e=.5*((e1+e2)-sqrt((e1-e2)**2+4.*h12**2))
            write(*,20) r,e1,e2,e
            r=r+0.05
        enddo
10      format(2x,'distance   e1    e2     Eg')
20      format(2x,f10.6,3(4x,f16.6))
        stop
        end
```

Data for 1.c

61. 0.92 2.27 313. 83. 20 134. 640

1.D.   VB potential Surface for the A − B + C → A + B − C Reaction

```
c       This program calculates the VB potential energy surface
c       for the three atoms four electron problem.
c       j           The integral obtained from Morse and anti-Morse functions.
c       k           The integral obtained from Morse and anti-Morse functions.
        real j(3),k(3),r(3),Eg(10,10)
        common/code/iac(3)
        print*,'input the bond codes for A-B, B-C and A-C respectively'
        print*,'(these codes are between 1 and 22 according to the sequence'
        print*,'in subroutine morse)'
        read*,(iac(i),i=1,3)
        do 5 ir=1,10
            do 5 jr=1,10
                r(1)=0.80+0.2*(jr-1)                          ! Adjusts the bond lengths.
                r(2)=1.00+0.2*(ir-1)
                r(3)=r(2)+r(1)
                do 10 m=1,3
                    call morse(r,k,j,m)
10              continue
                fkk=0.5*((k(1)-k(2))**2+(k(2)-k(3))**2+(k(3)-k(1))**2)

                Eg(ir,jr)=1./(1.+.18)*(j(1)+j(2)+j(3)-sqrt(fkk))    ! Evaluates the potential surface.
```

```
5          continue
           write(6,25)
25         format(5x,' Energy surface')
           write(6,20) (( Eg(ir,jr),jr=1,10),ir=1,10)
20         format(10(f7.2,1x))
           end
c===============================================================
           subroutine morse(r,kt,jt,m)

c     d          The bond dissociation energy
c     rz         The bond length
c     fm         The Morse function
c     fms        The anti Morse function
c     The iac code selects potential parameters according to the bond
c     type by assigning iac values between 1 to 22 to the following
c     bonds, H-H, O-O, N-N, C-C, F-F, Cl-Cl, Br-Br, I-I, S-S, O-H, C-H,
c     N-H, H-F, H-Cl, H-Br, H-I, H-S, C-O, C-F, C-Cl, C-Br, C-I
      real jt(3),kt(3),d(22),rz(22),aa(22)
      real dd(3),r0(3),fa(3),r(3)
      common/code/iac(3)
      data d/104.,54.,58.,88.,36.,58.,47.,36.,51.,102.,108.5,103.,134.,
     & 106.5,89.,74.,81.,92.,105.,81.,66.,54./
      data rz/0.74,1.32,1.40,1.54,1.42,1.99,2.28,2.67,2.08,0.96,1.09,
     & 1.00,0.92,1.28,1.41,1.60,1.32,1.43,1.39,1.78,1.94,2.14/
      data aa/1.99,2.32,2.30,1.60,2.98,2.03,1.94,1.85,1.83,2.26,1.81,
     & 2.07,2.27,1.85,1.82,1.75,1.87,1.87,1.81,1.72,1.58,1.56/
      dd(m)=d(iac(m))
      r0(m)=rz(iac(m))
      fa(m)=aa(iac(m))
      dr=r(m)-r0(m)
      ex1=exp(-fa(m)*dr)
      ex2=ex1*ex1
      fm=dd(m)*(ex2-2*ex1)
      fms=0.5*dd(m)*(ex2+2*ex1)
      jt(m)=0.5*(fm+fms+0.18*(fm-fms))
      kt(m)=0.5*(fm-fms+0.18*(fm+fms))
      return
      end
```

REFERENCES

1. D. A. McQuarrie, *Quantum Chemistry*, University Science Books, 1983.
2. J. A. Pople and D. L. Beveridge, *Approximate Molecular Orbital Theory*, McGraw-Hill, New York, 1970.
3. C. C. J. Roothaan, *Rev. Mod. Phys.*, **23**, 69 (1951).
4. H. Eyring, J. Walter, and G. E. Kimball, *Quantum Chemistry*, Wiley, New York, 1967.
5. C. A. Coulson and U. Danielsson, *Ark. Fys.*, **8**, 239 (1954).
6. K. B. Wiberg, *Physical Organic Chemistry*, Wiley, New York, 1963.

2

CHEMICAL REACTIONS IN THE GAS PHASE AND IN SIMPLE SOLVENT MODELS

2.1. REACTION RATE IN THE GAS PHASE

2.1.1. Equilibrium Constant and Rate Constants

In order to consider the relationship between potential surfaces and chemical reactivity we start by reviewing the relevant concepts. To do this we examine the reaction

$$A + B \underset{k_{-1}}{\overset{k_1}{\rightleftharpoons}} C + D \tag{2.1}$$

The rate of change of the concentration of A can be expressed as

$$d[A]/dt = -k_1[A][B] + k_{-1}[C][D] \tag{2.2}$$

The constants k_1 and k_{-1} are, respectively, the forward and backward *rate constants* and their ratio can be expressed by the *law of mass action* as

$$k_1/k_{-1} = ([C][D])_{eq}/([A][B])_{eq} = K_{eq} \tag{2.3}$$

where K_{eq} is the *equilibrium constant* and $(\ )_{eq}$ designates the value of the quantity in brackets at equilibrium. To explore the relationship between the reaction potential surface and the rate constant we start by considering a section of the potential surface of the $CH_4 + Cl \rightarrow CH_3 + HCl$ (Fig. 1.7) along its least-energy path which is taken as the *reaction coordinate*. This section shown in Fig. 2.1 is characterized by the height of the transition state and the corresponding barriers $\Delta U_1^{\neq}$ and $\Delta U_{-1}^{\neq}$, which are called the forward and backward *activation barriers*. Intuitively it appears that k_1 and k_{-1} are determined, respectively, by the $\Delta U_1^{\neq}$ and $\Delta U_{-1}^{\neq}$ activation barriers. In fact, one can guess a Boltzmann-like dependence of the rate constant on the corresponding barriers, that is,

$$k_1 \approx A \exp\{-\Delta U_1^{\neq}\beta\} \tag{2.4}$$

where $\beta = (k_B T)^{-1}$ and k_B is the Boltzmann constant. The reason for this is apparent from the figure; that is, as long as the kinetic energy along the

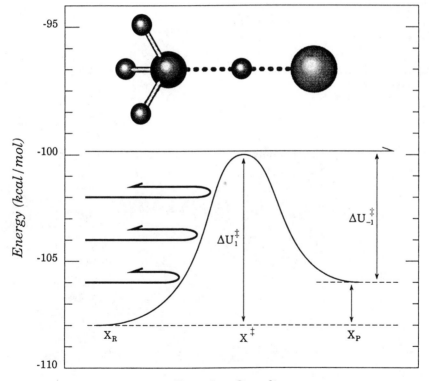

FIGURE 2.1. The least-energy path of the $CH_4 + Cl \rightarrow CH_3 + HCl$ reaction. This one-dimensional curve is obtained by cutting the two-dimensional surface of Figure 1.7 along the $a \rightarrow b \rightarrow c \rightarrow d \rightarrow e$ path, which is taken here as the reaction coordinate.

reaction coordinate is lower than $\Delta U_1^{\neq}$, the colliding $CH_4 + Cl$ fragments will be reflected back and stay on the reactant side of the barrier.

To evaluate the rate constant in a more rigorous way (as is done in detail in Ref. 1), let us consider first the average behavior of many reacting systems with a one-dimensional surface of the type described in Fig. 2.1. We will try to determine what fraction of the systems that pass the reactant region toward the barrier would react. The number of systems in a path length Δx, which have a momentum between p and $p + \Delta p$, is given by (see Ref. 1b)

$$C(p) = Z_R^{-1} \exp\left\{-\frac{p^2}{2m}\beta\right\} \frac{\Delta x\,\Delta p}{h} \tag{2.5}$$

where $Z_R = (\sqrt{2\pi m k_B T}/h) \int_{-\infty}^{x^{\neq}} \exp\{-U\beta\}\,dx$ is the so called *classical partition function* for the reactant (R) state. The number of such systems which pass a given point in the reaction path per second will be $C(p)p/(m\,\Delta x)$ [since $p/(m\,\Delta x)$ is the inverse time for passing the length Δx] and their contributions to the rate constant will be

$$\Delta k = \bar{F}(p)C(p)\frac{p}{m\,\Delta x} \tag{2.6}$$

where $\bar{F}$ is the *transmission factor* which expresses the fraction of systems that reach the transition state and react. Now k is obtained by integrating Δk and is given by

$$k = \int_0^{\infty} \bar{F}(p) \exp\left\{-\frac{p^2}{2m}\beta\right\} p\,\Delta p(mhZ_R)^{-1} \tag{2.7}$$

where we have considered only positive p. Using $\bar{F} = 0$ for $\frac{p^2}{2m} \le U^{\neq}$, and $\bar{F} = F$ for $\frac{p^2}{2m} > U^{\neq}$, we obtain

$$\begin{aligned}
k &= \bar{F}\int_{\sqrt{2U_1^{\neq}m}}^{\infty} \exp\left\{-\frac{p^2}{2m}\beta\right\} p\,\Delta p(mhZ_R)^{-1} \\
&= \bar{F}(k_B T/h)(e^{-U^{\neq}\beta}/Z_R) \\
&= \frac{1}{2}\bar{F}(2k_B T/\pi m)^{1/2}\left(e^{-U^{\neq}\beta}\Big/\int_{-\infty}^{x^{\neq}} e^{-U\beta}\,dx\right) \\
&= \frac{1}{2}F|\dot{x}|\left(e^{-U^{\neq}\beta}\Big/\int_{-\infty}^{x^{\neq}} e^{-U\beta}\,dx\right).
\end{aligned} \tag{2.8a}$$

where we use the fact that the factor $(2k_B T/\pi m)^{1/2}$ is the average of the absolute value of the velocity $\dot{x}$.

Our rate constant can also be expressed in the alternative form

$$k = \frac{1}{2}F\tau^{-1}\left(\Delta x^{\neq} e^{-U^{\neq}\beta}\Big/\int_{-\infty}^{x^{\neq}} e^{-U\beta}\,dx\right) = \frac{1}{2}F\tau^{-1}(z^{\neq}/z_R) \tag{2.8b}$$

where τ is the time needed to cross the transition state region, $\Delta x^{\neq}$, with a velocity $\dot{x}$ (this time will be used in subsequent sections).

The relatively simple expression of eq. (2.8) and related derivations (Ref. 1) provide useful insight about gas phase reactions, although more sophisticated treatments are needed to obtain the exact reaction cross section. Nevertheless, the general form of eq. (2.8) appears to provide a quite rigorous framework for describing reactions of molecules with more than a few degrees of freedom and in the condense phase (where the energy distribution between the different degrees of freedom follow the simple statistical mechanical distribution rules). The treatment of rate constants in many-dimensional systems in condensed phases will be considered below.

2.1.2. The Rate Constants for Many-Dimensional Systems

In considering the equivalent of eq. (2.8) for multidimensional systems we will start by defining the relevant reaction coordinate, X, and the probability, $P(X)$, that the system will be at different points, along X. The reaction coordinate can be taken rather arbitrarily as any well-defined parameter [e.g., $X = (r_{23} - r_{12})$ in Fig. 1.7]. Once X is selected we can obtain $P(X)$ by dividing the coordinate space into subsets according to the specific value of X and evaluating the one-dimensional function.

$$P(X)\,\Delta X = \int_{-\infty}^{\infty} d\mathbf{s} \int_{X-\Delta X/2}^{X+\Delta X/2} dX \exp\{-U\beta\} \Big/ \int_{-\infty}^{\infty} d\mathbf{s} \int_{-\infty}^{\infty} dX \exp\{-U\beta\}$$
(2.9)

where $\mathbf{s}$ is the subspace of $3n - 1$ coordinates orthogonal to X (e.g., if X is taken as $r_{23} - r_{12}$ in Fig. 1.7, then $\mathbf{s} = r_{12} + r_{23}$). This probability function defines the many-dimensional equivalent of the potential of Fig. 2.1 which is now expressed as a *free-energy function* or a *potential of mean force*

$$\Delta g(X) = -\beta^{-1} \ln[P(X)/P(X_R^0)]$$
(2.10)

where X_R^0 is the point that maximizes the $P(X)$ at the reactant state. The Δg of eq. (2.10) and the corresponding $P(X)$ can be used to evaluate the rate constant in a many-dimensional reactive system. That is, in cases of significant activation barriers, the reactant system spends most of its time encountering ineffective collisions. However, once in a long time the kinetic energy stored in the many degrees of freedom of the system is converted into a motion along the reaction coordinate, producing a *reactive trajectory* that reaches $X^{\neq}$. Thus, the rate constant can be expressed in terms of the probability that the system will be within a range $\Delta X^{\neq}$ around $X^{\neq}$ and the average time, $\bar{\tau}$, it takes a reactive trajectory to cross that region, as (Ref. 2):

$$k = \frac{\langle \dot{X}_+ \rangle P(X^{\neq})}{\int_{-\infty}^{x^{\neq}} P(x) \, dx} = \frac{\bar{\tau}^{-1}(P(X^{\neq}) \Delta X^{\neq})}{\int_{-\infty}^{X^{\neq}} P(X) \, dX}$$

$$= \bar{\tau}^{-1} \int_{X^{\neq}-\Delta X^{\neq}/2}^{X^{\neq}+\Delta X^{\neq}/2} \exp\{-\Delta g(X)\beta\} \, dX \Big/ \int_{-\infty}^{X^{\neq}} \exp\{-\Delta g(X)\beta\} \, dX$$

$$\approx \bar{\tau}^{-1} \exp\{-\Delta g(X^{\neq})\beta\} \left(\Delta X^{\neq} \Big/ \int_{-\infty}^{X^{\neq}} \exp\{-\Delta g(X)\beta\} \, dX \right)$$

$$= \bar{\tau}^{-1} \exp\{-\Delta G^{\neq}\beta\} \tag{2.11a}$$

where $\langle \dot{X}_+ \rangle$ is the average velocity along X for trajectories that arrive at $X^{\neq}$ from the reactant and continue to the product region without being deflected backward. The magnitude of $\Delta X^{\neq}$ is usually determined by the relationship $([\Delta g(X^{\neq}) - \Delta g(X^{\neq} - \Delta X^{\neq}/2)] = \beta^{-1})$, but different values of $\Delta X^{\neq}$ can also be used (this will give different values of $\bar{\tau}^{-1}$ and $\Delta G^{\neq}$ but a similar value of k).

In deriving eq. (2.11) we introduced the notation $\Delta G^{\neq}$, which can also be expressed as

$$\exp\{-\Delta G^{\neq}\beta\} = \Delta X^{\neq} P(X^{\neq}) \Big/ \int_{-\infty}^{X^{\neq}} P(X) \, dX = z^{\neq}/z_R \tag{2.11b}$$

where z is the configurational part of the classical partition function, Z, that has been introduced in the one dimensional case of eq. (2.5). Note the distinction between $\Delta G^{\neq}$ and $\Delta g^{\neq}$ when these terms are used in subsequent chapters. It is also important to note that the time $\bar{\tau}$ is not much different from the time it takes a productive trajectory to move from the reactant region to the transition state. The value of $\bar{\tau}^{-1}$ is similar for many types of reactions in condensed phases and can be approximated at room temperature by

$$\bar{\tau}^{-1} = \left[\frac{\langle \dot{X}_+ \rangle^{\neq}}{\Delta X^{\neq}} \right] \approx 10^{13} \text{ sec}^{-1} \tag{2.12}$$

where $\Delta X^{\neq}$ is of the order of 1 Å.

Equation (2.11) is frequently expressed in the form of eq. (2.8) as

$$k = \frac{1}{2} F \langle |\dot{X}| \rangle P(X^{\neq}) \Big/ \int_{-\infty}^{X^{\neq}} P(X) \, dX = \frac{1}{2} F(\langle |\dot{X}| \rangle / \Delta X^{\neq}) z^{\neq}/z_R \tag{2.13}$$

Here, as before, the transmission factor F expresses the fraction of the trajectories which continue to the product state after arriving at $X^{\neq}$ (see Fig.

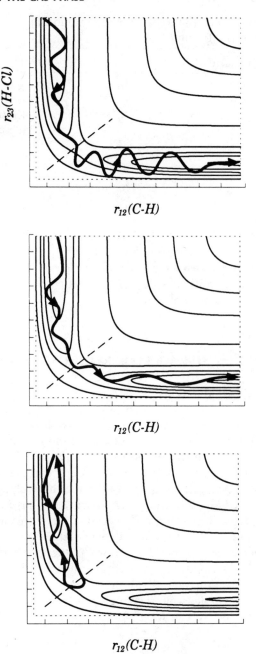

FIGURE 2.2. A schematic description of the evaluation of the transmission factor F. The figure describes three trajectories that reach the transition state region (in reality we will need many more trajectories for meaningful statistics). Two of our trajectories continue to the product region X_P, while one trajectory crosses the line where $X = X^{\neq}$ (the dashed line) but then bounces back to the reactants region X_R. Thus, the transmission factor for this case is 2/3.

TABLE 2.1.   The approximated relationship between $\Delta g^{\neq}$ and k^{-1}

$\Delta g^{\neq}$ (kcal/mol)	k^{-1} (sec)	Typical Times
0–2	10^{-12}	picoseconds
5	10^{-9}	nanoseconds
10	10^{-6}	microseconds
15	10^{-3}	milliseconds
18	10^{0}	seconds

2.2). When F is equal to unity, the equation reduces to the rate expression of the well-known *transition state theory*. In most of the cases considered in this book, we will deal with reactions in condensed phases where F is not much different from unity and the relation between k and $\Delta g^{\neq}$ follows the qualitative role given in Table 2.1.

2.2. VB POTENTIAL SURFACES FOR REACTIONS IN SOLUTIONS

2.2.1. General Considerations

Extending the considerations discussed above to chemical reactions in solutions requires one to evaluate the effect of the solvent on both $\Delta g^{\neq}$ and $\bar{\tau}$. While the effect of the solvent on $\bar{\tau}$ is not a major one (this effect will be considered in Chapter 3), its effect on k through the change of $\Delta g^{\neq}$ can be enormous. For example, let us consider the heterolytic bond cleavage $X–Y \rightarrow X^+ + Y^-$ of Fig. 2.3 in a polar solvent. As illustrated in the figure, the potential surface in solution is drastically different from the corresponding gas phase potential. The excited ionic state is now stabilized by the solvent and becomes a ground state when the bond is broken to two solvated ions. The simplest way of treating this effect by the VB formalism is to add the solvation energy of the ionic state to the corresponding diagonal matrix element, while leaving the off-diagonal element in its gas phase value. That is, we can write

$$\varepsilon_1 = H_{11} = H_{11}^0$$

$$\varepsilon_2 = H_{22} = H_{22}^0 + \frac{1}{2} U_{Ss}^{(2)}$$

$$H_{12} = H_{12}^0 \qquad (2.14)$$

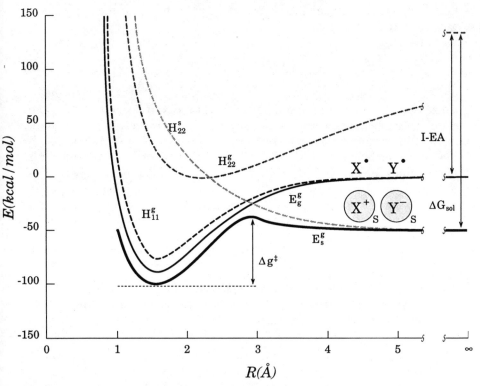

FIGURE 2.3. The energetics of a heterolytic bond cleavage reaction in a polar solvent. The specific example shown corresponds to the $CH_3OCH_3 \rightarrow CH_3^+ + CH_3O^-$ reaction in water. The energy of the covalent state does not include the effect of the solvent on this state, but a more consistent treatment (e.g., eq. (2.21) should account for the polarization of the solvent toward the charges of the ionic state. This would result in *destabilization* of H_{11}^s.

where $U_{Ss}^{(2)}$ is the solute–solvent interaction potential of the ionic state (the factor half reflects the assumption that half of the energy gained from polarizing the solvent by the ionic state is invested on solvent–solvent interactions). This is basically the approach we took in treating the interaction of the H_2 molecule with an external charge (Chapter 1), but now we deal with a solvated state. The present case, however, involves many solvent configurations where we should consider the free energy, $\Delta g(R)$, that corresponds to the ground state surface obtained by mixing H_{11} and H_{22} at different solvent configurations. The simplest way to accomplish this task is to consider the average value of H_{22} over all the configurations of the solvent

$$H_{22} \approx \Delta g^{(2)} = H_{22}^0 + \Delta g_{sol}^{(2)} \tag{2.15}$$

(where $\Delta g_{sol}^{(i)}$ is the solvation free energy associated with $U_{Ss}^{(i)}$), and then to take the ground state obtained by mixing $\Delta g^{(2)}$ with H_{11} as the approximate ground-state free energy.

$$\Delta g(R) \approx \frac{1}{2} [(H_{11} + \Delta g^{(2)}) - ((H_{11} - \Delta g^{(2)})^2 + 4H_{12}^2)^{1/2}] \qquad (2.16)$$

Thus, our task is reduced to estimating the solvation free energy of the ionic state. This solvation energy should be estimated, however, using the charge distribution of the ground state obtained by mixing ψ_1 and ψ_2, rather than by the pure ionic state ψ_2 (since the solvent configurations are determined by the ground-state charges). The simplest way to estimate this energy is to use a continuum model where the solvation energy of the dipolar state is given (in kcal/mol) by the product of the field from the solvent, ξ, and the solute dipole $\mu^{(2)}$,

$$\Delta g_{sol}^{(2)} = -166\xi^{(g)} \cdot \mu^{(2)} \qquad (2.17)$$

The field $\xi^{(g)}$ is the so-called "reaction field" which is due in the present case to the ground state dipole of the solute, $\mu^{(g)}$, and is given (in $\mathring{A}^{-2}$ and electron charge units) by (Ref. 3)

$$\xi^{(g)} = 2(\mu^{(g)}/\bar{a}^3)((d-1)/(2d+1)) \qquad (2.18)$$

Here, $\bar{a}$ is the radius of the cavity around the solute (given in $\mathring{A}$), the dipole μ is given in $\mathring{A}$ and au, and d is the macroscopic dielectric constant of the solvent. The crucial problem, however, is that the cavity radius is an arbitrary parameter which is not given by the macroscopic model, making the results of eq. (2.18) rather meaningless from a quantitative point of view. A much more quantitative model is provided by the semimicroscopic model described below.

Exercise 2.1. Evaluate the ground-state potential surface for the $CH_3OCH_3 \rightarrow CH_3^+ + CH_3O^-$ reaction using the reaction field model, with a cavity radius $\bar{a} = R/2 + 1.5$.

Solution 2.1. The parameters and functions that should be used for the gas-phase matrix elements are given in Chapter 1 [eqs. (1.57) and (1.59) and Table 1.1]. Thus the only additional step for the evaluation of eq. (2.16) is the estimate of $\Delta g_{sol}^{(2)}$ of eqs. (2.17) and (2.18). Usually $\mu^{(g)}$ is a function of $\Delta g_{sol}^{(2)}$ (see subsequent sections), but here we use for simplicity $\mu^{(g)} = \mu^{(2)}(1 - \exp\{-R\})$. After calculating $\Delta g_{sol}^{(2)}$ and the corresponding $\Delta g(R)$ for several values of R and trying to construct the equivalent of Fig. 2.3, you will find your results to be very disappointing (the correct results are given in Fig. 2.5). This will be the case for other reasonable choices of the cavity radius. Thus, we conclude that the macroscopic estimate of eq. (2.18) *should not* be used in quantitative studies of chemical reactions in solutions.

To restate the conclusion from the above exercise and related studies; regardless of the wide appeal of the reaction field formulation (i.e., the elegant study of Ref. 8) it should not be used in realistic studies of chemical reactions in solutions. The reason for this is quite simple, a method for calculation of charge transfer or charge separation in solutions is as good as its ability to evaluate solvation energies of the relevant fragments. An approach that cannot give reliable solvation energy will not give quantitative results, regardless of the accuracy of the quantum mechanical method used to evaluate the gas phase Hamiltonian matrix elements.

2.2.2. The Langevin Dipoles Model

The evaluation of the solvation energies of ionic states presents a major challenge; the traditional macroscopic models (Ref. 3) depend on the unknown cavity radius of the solute [a in eq. (2.18)]. Fully microscopic models, on the other hand, require us to include explicitly a huge number of solvent molecules and involve major convergence problems. An alternative approach that overcomes some of the convergence problems of the fully microscopic model, yet treats the solvent explicitly in a simplified way, is the Langevin dipoles model (Ref. 4). This model represents the time average polarization of the solvent molecules by a cubic grid of polarizable dipoles. The model is constructed in three steps (Fig. 2.4). (*a*) A cubic grid (typically with 3 Å spacing) is placed around the solute atoms. (*b*) Each grid point which is within a van der Waals distance, r^*, from a solute atom is excluded. (*c*) The remaining grid points are then replaced by point dipoles whose polarization should mimic the average polarization of the solvent molecules at the same region in space. This is accomplished by using a Langevin-type relationship,

$$\boldsymbol{\mu}_i^{n+1} = [\coth(X_i^n) - 1/X_i^n]\mu_0 \boldsymbol{\xi}_i^n / \xi_i^n$$

$$X^n = C\mu_0 \xi^n / k_B T$$

$$\boldsymbol{\xi}_i^n = \boldsymbol{\xi}_i^0 + \boldsymbol{\xi}_{\mu,i}^n \tag{2.19}$$

where $\boldsymbol{\xi}_i^0$ is the field on the ith dipole from the solute charges and $\boldsymbol{\xi}_{\mu,i}$ is the field on the ith dipole from all other dipoles. The index n indicates that we are dealing with an iterative procedure starting from $\boldsymbol{\xi}_{\mu,i}^0 = 0$. The parameters in this model (C and μ_0) can be obtained (Ref. 4) by using an explicit all-atom solvent model and molecular dynamics simulation to evaluate the field-dependent polarization of water molecules around ions and then fitting

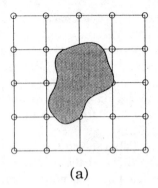

(a)

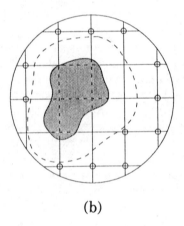

(b)

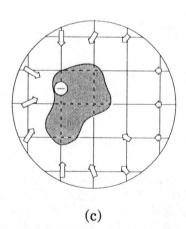

(c)

FIGURE 2.4. A schematic description of the Langevin Dipole model. The figure illustrates the three steps involved in constructing the model.

TABLE 2.2. Parameters for the LD Model

Parameter[a]	Value
μ_0	0.35
C	1.00
r_O^*	2.40
$r_{O^-}^*$	2.60
r_H^*	1.80
r_N^*	3.00
r_C^*	3.00

[a]μ_0 is given in Å electron charge units, r^* is the closest distance (in Å) between the grid points to the indicated solute atom.

eq. (2.19) to the corresponding results of the simulation. The free energy of the Langevin dipoles (LD) model is simply given (in kcal/mol) by (Ref. 4)

$$\Delta g_{sol}^n = -166 \sum_i \mu_i^n \xi_i^0 \tag{2.20}$$

where the units used are the same as those used in eq. (2.17). This expression reflects the fact that half of the energy gained from a dipole-field interaction is spent on polarizing the dipoles. Note that the dipole μ^n reflects both the fields ξ^0 and ξ_μ (see Ref. 4). Finally, the parameters of the LD model are given in Table 2.2.

Exercise 2.2. Implement the Langevin Dipole method in a computer program. Use this model to evaluate the solvation energy of the CH_3O^- ion.

Solution 2.2. See Program 2.A.

2.2.3. LD Calculations of VB Potential Surfaces

Although the LD model is clearly a rough approximation, it seems to capture the main physics of polar solvents. This model overcomes the key problems associated with the macroscopic model of eq. (2.18), eliminating the dependence of the results on an ill-defined cavity radius and the need to use a "dielectric constant" which is not defined properly at a short distance from the solute. The LD model provides an effective estimate of solvation energies of the ionic states and allows one to explore the energetics of chemical reactions in polar solvents.

In order to clarify the implementation of the LD model in VB calculations, it is useful to apply this model to the heterolytic bond cleavage of

Section 2.2.1. Our previous attempt to evaluate the relevant potential surface using eq. (2.17) encountered a major difficulty in trying to estimate the solvation energy of the ionic state $\Delta g_{sol}^{(2)}$. Now with the LD model we can try to obtain a reliable estimate of the ground-state free energy $\Delta g(R)$ of eq. (2.16). To accomplish this we need to evaluate the solvation energy of the ionic state ψ_2 as a function of R. This is done by fixing R and building a grid around the given solute (see Program 2.A). After excluding grid points which are within a van der Waals distance from the solute atoms, we evaluate the interaction between the ionic state and the self-consistently polarized Langevin dipoles, situated at the remaining grid points. Now, however, we must introduce a complication; the Langevin dipoles should be polarized by the charge distribution of the ground state obtained by mixing H_{11} and H_{22}, rather than by the purely ionic charge distribution of ψ_2. Thus we solve the secular equation $\mathbf{HC} = \mathbf{EC}$ in an iterative way using

$$H_{11}^{(n+1)} = H_{11}^0 - \frac{1}{2} \sum_i Q_i^{(n)} V_i^{(n)} \simeq H_{11}^0 + \Delta g_{sol,n}^{(1)}$$

$$H_{22}^{(n+1)} = H_{22}^0 + \sum_i q_i^{(2)} V_i^{(n)} - \frac{1}{2} \sum_i Q_i^{(n)} V_i^{(n)} \simeq H_{22}^0 + \Delta g_{sol,n}^{(2)}$$

$$Q_i^{(n)} = (C_{g2}^{(n)})^2 q_i^{(2)}$$

$$V_i^{(n)} = \sum_j \mathbf{r}_{ji} \mu_j(\mathbf{Q}^{(n)}) / r_{ij}^3 \tag{2.21}$$

where $\mathbf{r}_{ji} = \mathbf{r}_i - \mathbf{r}_j$ and $V_i^{(n)}$ is the potential on the ith solute atom from the solvent dipoles in their nth iteration, $\mathbf{q}^{(2)}$ and $\mathbf{Q}$ are, respectively, the atomic charges of ψ_2 and of the ground state; $\mu(\mathbf{Q}^{(n)})$ designates the solvent dipoles [polarized by the given $\mathbf{Q}^{(n)}$]; $\mathbf{C}_g^{(n)}$ is the ground-state eigenvector of the secular equation obtained with $H_{22}^{(n)}$ and $C_{g2}^{(n)}$ is the 2nd component of this vector. The term $(-\frac{1}{2} \sum Q^{(n)} V^{(n)})$ reflects the energy invested in polarizing the solvent dipoles. Once we obtain converging results for H_{22}, we use the ground state of the $\mathbf{H}$ matrix [as given by eq. (2.16)] as our ground-state free energy. The best way to follow this discussion is by performing the following exercise and examining its solution.

Exercise 2.3. Use the LD model and eq. (2.21) to evaluate the potential energy surface for the $CH_3OCH_3 \rightarrow CH_3^+ + CH_3O^-$ reaction in water. Treat the two fragments as two effective atoms whose solvation free energy is similar to that of the actual fragments. Typical, solvation energies are given in Table 2.3.

Solution 2.3. Use Program 2.B (the corresponding results are plotted in Fig. 2.5).

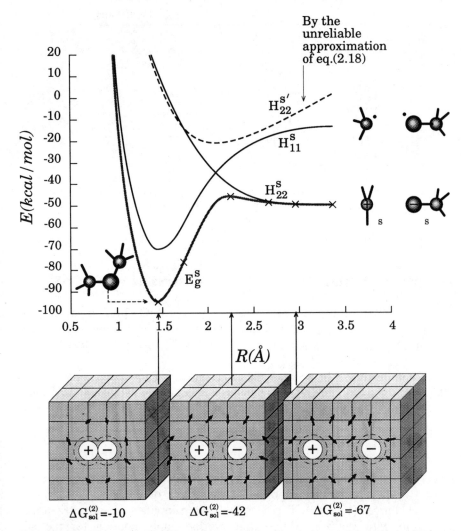

FIGURE 2.5. The results of LD calculations of the dissociation of CH_3OCH_3 to CH_3^+ and CH_3O^- in water. The figure shows H_{11} and H_{22} and the corresponding ground state E_g obtained from their mixing. The lower part of the figure illustrates schematically the change in the LD polarization upon increase in the C–O bond length (R) and gives the corresponding calculated solvation energies of the ionic state. The figure also illustrates the very unreliable result obtained by using eq. (2.18). The solvation energy of the covalent state is given for simplicity with the same approximation used in Fig. 2.3 rather than eq. (2.21).

TABLE 2.3. Solvation Free Energies of Typical Ions[a]

Ion	$\Delta G_{sol,w}$ (kcal/mol)
Cl^-	-75
Na^+	-102
Ca^{2+}	-380
OH^-	-110
H_3O^+	-110
$HCOO^-$	-80
CH_3O^-	-95
$CH_3NH_3^+$	-75
Imidazole$^+$	-62
CH_3^+	-90

[a] $\Delta G_{sol,w}$ designates the solvation free energy of the given ion in water. The corresponding observed information is compiled and analyzed in Ref. 4.

2.3. MO POTENTIAL SURFACES FOR REACTIONS IN SOLUTIONS

While the VB approach is more effective than the MO approach in treating reactions in solutions, it is useful to be familiar with both approaches. Here we outline a simple and general procedure that incorporates solvent effects into MO calculations.

As demonstrated in Section 1.2, it is quite simple to incorporate the effect of a single external charge in MO calculations. The same approach can be extended to a collection of external charges writing

$$H_{\nu\nu} \approx H_{\nu\nu}^0 - \sum_C Q_C r_{\nu C}^{-1} \qquad (2.22)$$

where $\mathbf{H}^0$ is the H matrix of the reacting molecule in a vacuum and the second term represents the potential at the site of the atom ν from the external charges (the Q's). This seemingly oversimplified treatment is in fact consistent with the zero differential overlap approximation for the all-valence electron MO problem. Since eq. (2.22) is quite general it can be used with the charge distribution of any solvent at any given configuration. Thus, we can use the dipole approximation and write

$$H_{\nu\nu} \approx H_{\nu\nu}^0 - \sum_C \boldsymbol{\mu}_C \mathbf{r}_{C\nu} / r_{C\nu}^3 \qquad (2.23)$$

where $\mathbf{r}_{C\nu} = \mathbf{r}_\nu - \mathbf{r}_C$ and the second term reflects the potential from the solvent dipoles at the site of the solute atom. The total energy is now given by

$$E = E^0 + \sum_A \sum_C Q_A V_{CA} \tag{2.24}$$

where V_{CA} is the potential from the Cth solvent dipole at the site of the Ath solute atom.

In general we have to use eq. (2.24) iteratively, starting with the gas-phase charge distribution of the solute, evaluating the polarization of the solvent dipoles in the presence of these charges, substituting these dipoles in eq. (2.23), and obtaining new $\mathbf{H}$ and new charges, which are then used to obtain new solvent polarization. This can be expressed by

$$H_{\nu\nu}^{(n)} = H_{\nu\nu}^0 - \sum_C \boldsymbol{\mu}_C(\mathbf{Q}^{(n-1)})\mathbf{r}_{C\nu}/r_{C\nu}^3 = H_{\nu\nu}^0 - V_\nu^{(n)} \tag{2.25}$$

where $V_\nu^{(n)}$ is the potential from the solvent, in the nth iteration, at the site of the νth solute atom. This procedure converges much more slowly than eq. (2.21).

Exercise 2.4. Repeat Exercise 1.6 with the LD solvent model instead of the external charge.

Solution 2.4. Evaluate the V_ν's of Eq. (2.25) using Program 2.C (these V's are called Vq in the program).

2.4. PROTON TRANSFER REACTIONS AND THE EVB MODEL

2.4.1. VB Potential Surface for Proton Transfer Reactions in Solutions

A very useful example of treating solvent effects is provided by considering the proton transfer reaction

$$RXH + Y \rightarrow RX^- + HY^+ \tag{2.26}$$

This reaction can be described by the three resonance structures

$$\phi_1 = RX \overset{\cdot\cdot}{-\!-} H \qquad Y:$$

$$\phi_2 = RX:^- \qquad H \overset{\cdot\cdot}{-\!-} Y^+$$

$$\phi_3 = RX:^- \qquad H^+ \qquad Y: \tag{2.27}$$

The electrons involved in the actual reaction (which are designated here by dots and referred to here as the *active electrons*) can be treated according to the general prescription of the four-electron three-orbital problem with the VB wave functions (Ref. 5)

$$\Phi_1 = N_1\{|X\bar{H}Y\bar{Y}| - |\bar{X}HY\bar{Y}|\}\chi_1 = \phi_1\chi_1$$

$$\Phi_2 = N_2\{|X\bar{X}H\bar{Y}| - |X\bar{X}\bar{H}Y|\}\chi_2 = \phi_2\chi_2$$

$$\Phi_3 = N_3|X\bar{X}Y\bar{Y}|\chi_3 \qquad\qquad = \phi_3\chi_3 \qquad (2.28)$$

where X, Y, and H designate atomic orbitals on the corresponding atoms. The N's are normalization constants and the χ's are the wave functions of the inactive electrons moving in the field of the active electrons. The three resonance structures Φ_1, Φ_2, and Φ_3 can be treated by the approach detailed in Appendix B and can be reduced to an effective two-state problem, where one state is mostly Φ_1 and the other is mostly Φ_2. The corresponding matrix elements can be evaluated by standard quantum chemical methods, but this evaluation is very tedious. Instead, we can exploit the simple physical picture of Φ_1 and Φ_2 and describe H_{11} and H_{22} by simple potential functions that can be calibrated by both experimental information and accurate quantum mechanical calculation. That is, the function H_{11} will be given, at the range where the X–Y distance is large as compared to the X–H bond length, by a Morse potential function that depends on the distance R_{X-H}. When the H atom approaches Y, we have a repulsive van der Waals interaction between these atoms. We can describe both of these forces using analytical potential energy terms (see discussion below). The same argument applies to H_{22}. As far as H_{12} is concerned, we can approximate it by an exponential term and fit the parameters in this term to the experimental information on the gas-phase potential energy surface of the reaction or, if needed, to accurate gas-phase calculations.

Thus, we describe the gas-phase potential by

$$\varepsilon_1^0 = H_{11}^0 = \Delta M(b_1) + U_{nb}^{(1)} + (K/2)(\theta_1 - \theta_1^0)^2 + U_{inact}^{(1)}$$

$$\varepsilon_2^0 = H_{22}^0 = \Delta M(b_2) + U_{nb}^{(2)} - 332/r_3 + \alpha_2^0 + U_{inact}^{(2)}$$

$$H_{12}^0 = A \exp\{-\mu(r_3 - r_3^0)\} \qquad (2.29)$$

where b_1, b_2, and r_3 are, respectively, the X–H, H–Y, and X–Y distances, θ_1 is the R–X–H bond angle, ΔM is a Morse potential function taken relative to its minimum value ($\Delta M(b) = M(b) - D$), and $U_{nb}^{(i)}$ is the repulsive nonbonded interaction in the given configuration. The parameter α_2^0 expresses the difference between the asymptotic energies of ϕ_1 and ϕ_2 (the energies evaluated at the configurations where the fragments of each resonance structure are at infinite separation). The potentials $U_{inact}^{(i)}$ represent the interaction within the *inactive* region of the reacting system and also the interaction between the active and inactive regions (see Appendix B). These potentials are described by

$$U_{inact}^{(i)} = 1/2 \sum_m K_{b,m}^{(i)}(b - b_{0,m}^{(i)})^2$$

$$+ 1/2 \sum_m K_{\theta,m}^{(i)}(\theta_m - \theta_{0,m}^{(i)})^2 + U_{nb,inact}^{(i)} + U_{nb,inact-act}^{(i)} \qquad (2.30)$$

where the b's and θ's are, respectively, the bond lengths and bond angles in both the inactive region and the connection between the active and inactive regions. The K_b, K_θ, b_0 and θ_0 are force field parameters which will be discussed in Chapter 4, while the U_{nb} terms represent the nonbonded interactions in the indicated regions.

The effect of the solvent on our reaction Hamiltonian is obtained by using the approach formulated in eq. (2.21), writing

$$\varepsilon_1^S = H_{11} = H_{11}^0 + \Delta g_{sol}^{(1)}$$

$$\varepsilon_2^S = H_{22} = H_{22}^0 + \Delta g_{sol}^{(2)}$$

$$H_{12} = H_{12}^0 \tag{2.31}$$

where $\Delta g_{sol}^{(i)}$ is the solvation energy of the ith state obtained by the iterative approach of eq. (2.21). The ground-state potential surface for the reaction is then obtained by solving the secular equations $\mathbf{HC} = E_g\mathbf{C}$. A typical surface for proton transfer in water is shown schematically in Fig. 2.6.

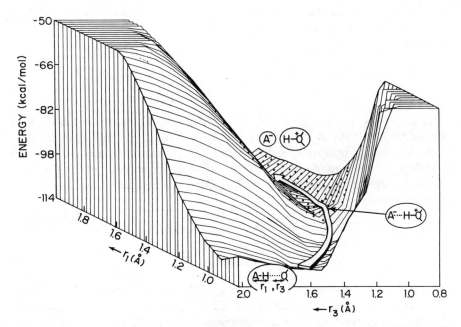

FIGURE 2.6. An EVB–LD potential surface for proton transfer between an acid R''COO$_A$H and an RO$_B$R' molecule in solution. The independent coordinates r_1 and r_3 are the distances between the proton and O$_A$ and O$_B$, respectively. Regions of the potential surface that have more than 50% ionic character are dotted (see Ref. 6 for more details).

2.4.2.   The EVB Calibration Procedure

The approach presented above is referred to as the *empirical valence bond* (EVB) method (Ref. 6). This approach exploits the simple physical picture of the VB model which allows for a convenient representation of the diagonal matrix elements by classical force fields and convenient incorporation of realistic solvent models in the solute Hamiltonian. A key point about the EVB method is its unique *calibration* using well-defined experimental information. That is, after evaluating the free-energy surface with the initial parameter α_2^0, we can use conveniently the fact that the free energy of the proton transfer reaction is given by

$$\Delta G_{PT} = \Delta G[(A - H + B) \rightarrow (A^- + HB^+)]$$

$$= 2.3RT[pK_a(A - H) - pK_a(B^+ - H)] \tag{2.32}$$

Now we can adjust α_2^0 until the calculated and observed ΔG_{PT} coincide. This calibrated surface can then be used with confidence for studying the reaction in different solvents (and different environments), since α_2^0 remains unchanged and only $\Delta g_{sol}^{(i)}$ is recalculated. In this way, the error associated with the evaluation of H_{ii}^0 does not affect the calculations of the relative effects of different solvents.

In general, when one deals with a more complicated reaction, for which it is hard to obtain gas phase estimates of α_i^0, it is convenient to use solution experiments from aqueous solutions to obtain the first estimate of α_i^0. This is done by using

$$\varepsilon_i(\infty) - \varepsilon_1(\infty) \simeq \alpha_i^0 + (\Delta g_{sol,w}^{i,\infty} - \Delta g_{sol,w}^{1,\infty}) = \alpha_i^0 + \Delta\Delta g_{sol,w}^{i,\infty} \simeq (\Delta G_{i,w}^\infty)_{obs} \tag{2.33}$$

where $\varepsilon_i(\infty)$ is the energy of the ith resonance structure when the relevant fragments are held at infinite separation and $\Delta g_{sol,w}^{i,\infty}$ is the corresponding solvation energy in water. Similarly $(\Delta G_{i,w}^\infty)_{obs}$ is the free energy involved in forming the ith configuration from the first configuration where the fragments in each configuration are held at infinite separation. This leads to the useful estimate

$$\alpha_i^0 \simeq (\Delta G_{i,w}^\infty)_{obs} - \Delta\Delta g_{sol,w}^{i,\infty} \tag{2.34a}$$

$$\Delta g_{sol,w}^{i,\infty} = \sum_k \Delta G_{sol,w}^{i,k} \tag{2.34b}$$

where $\Delta G_{sol,w}^{i,k}$ is the solvation energy of the kth fragment of the ith resonance structure, which can be easily estimated from observed solvation energies (e.g., the information compiled in Table 2.3) or by LD calculations. Since it is quite simple to obtain reliable estimates of the $\Delta G_{sol,w}$ and the corresponding $\Delta g_{sol,w}^{i,\infty}$, we can start with the estimate of eq. (2.34). In the next step we can calculate the actual $\Delta G_{i,w}^{\infty}$ of the different configurations and refine the α_i^0 until it reproduces the corresponding observed value.

The EVB approach described in this chapter provides a convenient way for estimating the energetics of chemical reactions in various solvents. However, the approximation involved in eq. (2.21) cannot be justified without detailed studies by more rigorous models. Such models will be described in Chapter 3.

2.5.   APPENDIX B—THE FOUR ELECTRONS/THREE ORBITALS VB TREATMENT

Many reactions can be described qualitatively by considering only three orbitals with four electrons. This includes proton transfer reactions $(A - H + B^- \rightarrow A^- + H - B)$ and S_N2 reactions $(X - Z + Y^- \rightarrow X^- + Z - Y)$. The VB description of these reactions requires in principle the six states depicted in Fig. 2.7, which considered the typical case of the $X^- + CH_3-Y \rightarrow X-CH_3 + Y^-$ S_N2 reaction.

However, a qualitative consideration of the relevant energetics (Table

TABLE 2.4.   "Back of the Envelope" Estimation of the Energies of Valence-Bond States of the $X^- + CH_3X \rightarrow XCH_3 + X^-$ S_N2 reaction[a]

Parameter	Value[b]	Abbreviation
Ionization potential of X	250	IP_X
Electron affinity of X	70	EA_X
Ionization potential of CH_3	230	IP_{CH_3}
Electron affinity of CH_3	$\simeq 0$	EA_{CH_3}
Carbon–halogen covalent bond	60	D_{C-X}
4.0-Å halogen–halogen bond	0	D_{C-X}
$+--$ Charge distribution energy	-60	$U_{QQ}(+--)$
$-+-$ Charge distribution energy	-180	$U_{QQ}(-+-)$

Approximate Valence–Bond State Energies

$E(1) =$ reference state energy $= 0$
$E(2) =$ reference state energy $= 0$
$E(3) = U_{QQ}(-+-) + IP_{CH_3} - EA_X + D_{C-X} = +40$
$E(4) = U_{QQ}(+--) + IP_X - EA_{CH_3} + D_{C-X} = +250$
$E(5) = U_{QQ}(+--) + IP_X - EA_{CH_3} + D_{C-X} = +250$
$E(6) = IP_X - EA_{CH_3} + D_{C-X} - D_{X-X} = +310$

[a] A rough estimate of the gas phase energy (in kcal/mol) of the states described in Fig. 2.7. The X–C and C–X' distances are both 2.0 Å and the X–X' distance is 4 Å. The notation X designates an average halogen.
[b] The sources of the various energy values are listed in Ref. 7.

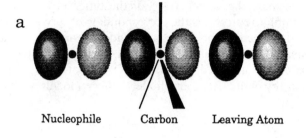

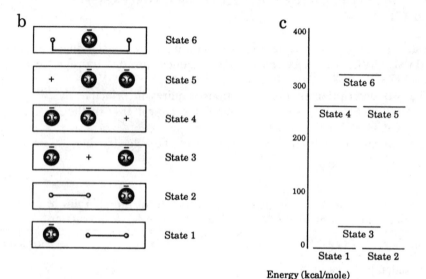

FIGURE 2.7. (*a*) Three active p_z orbitals that are used in the quantum treatment of the $X:^- + CH_3-Y \rightarrow X-CH_3 + Y:^-$ S_N2 reaction. (*b*) Valence-bond diagrams for the six possible valence-bond states for four electrons in three active orbitals. (*c*) Relative approximate energy levels of the valence-bond states in the gas phase (see Table 2.4 for the estimation of these energies).

2.4) indicates that the reaction can be described by three resonance structures

$$\Phi_1 = RX \div Z \quad Y:^-$$

$$\Phi_2 = RX:^- \quad Z \div Y$$

$$\Phi_3 = RX:^- \quad Z^+ \quad Y:^- \tag{2.35}$$

The electrons involved in the actual reaction, referred to here as the *active electrons*, can be treated with the VB wave functions [Refs. 5 and 7]:

$$\Phi_1 = N_1\{|X\bar{Z}Y\bar{Y}| - |\bar{X}ZY\bar{Y}|\}\chi_1 = \phi_1\chi_1$$

$$\Phi_2 = N_2\{|X\bar{X}Z\bar{Y}| - |X\bar{X}\bar{Z}Y|\}\chi_2 = \phi_2\chi_2$$

$$\Phi_3 = N_3|X\bar{X}Y\bar{Y}|\chi_3 \qquad\qquad = \phi_3\chi_3 \qquad\qquad (2.36)$$

where X, Y, and Z are the atomic orbitals on the X, Y, and Z atoms, respectively, while the N's are normalization constants and the χ's are the wave functions of the inactive electrons moving in the field of the active electrons. This requires that the Hamiltonian be of the form $\mathbf{H} = \mathbf{H}_{act} + \mathbf{H}_{inact} + \mathbf{H}'$ where $\mathbf{H}'$ represents the interaction between the subspaces of the active and inactive orbitals and the matrix elements between the basis wavefunctions are then given by

$$H_{ii} = \langle\Phi_i|\mathbf{H}|\Phi_i\rangle = \langle\phi_i|\mathbf{H}_{act}|\phi_i\rangle + \langle\chi_i|\mathbf{H}_{inact}|\chi_i\rangle + H'_{ii}$$

$$H_{ij} = \langle\Phi_i|\mathbf{H}|\Phi_j\rangle = \langle\phi_i|\mathbf{H}_{act}|\phi_j\rangle + \langle\chi_i|\mathbf{H}_{inact}|\chi_j\rangle + H'_{ij} \qquad (2.37)$$

The active electrons matrix elements are given by (Ref. 5)

$$\langle\phi_1|\mathbf{H}_{act}|\phi_1\rangle = (1 + S_1^2 - S_2^2)^{-1}(\langle 1|\mathbf{H}_{act}|1\rangle + \langle 1|\mathbf{H}_{act}|2\rangle - \langle 1|\mathbf{H}_{act}|3\rangle)$$

$$\langle\phi_2|\mathbf{H}_{act}|\phi_2\rangle = (1 - S_1^2 + S_2^2)^{-1}(\langle 5|\mathbf{H}_{act}|5\rangle - \langle 5|\mathbf{H}_{act}|6\rangle + \langle 5|\mathbf{H}_{act}|7\rangle)$$

$$\langle\phi_3|\mathbf{H}_{act}|\phi_3\rangle = \langle 4|\mathbf{H}_{act}|4\rangle$$

$$\langle\phi_1|\mathbf{H}_{act}|\phi_3\rangle = (2/(1 + S_1^2 + S_2^2))^{1/2}\langle 1|\mathbf{H}_{act}|4\rangle$$

$$\langle\phi_3|\mathbf{H}_{act}|\phi_2\rangle = (2/(1 - S_1^2 + S_2^2))^{1/2}\langle 4|\mathbf{H}_{act}|7\rangle$$

$$\langle\phi_1(\mathbf{H}_{act}|\phi_2\rangle = (2/(1 - (S_1^2 - S_2^2)^2))^{1/2}\langle 1|\mathbf{H}_{act}|5\rangle \qquad (2.38)$$

where $\quad 1 \equiv X(1)Z(2)Y(3)Y(4), \quad 2 \equiv Z(1)X(2)Y(3)Y(4), \quad 3 \equiv X(1)Y(2)$
$Y(3)Z(4), \quad 4 \equiv X(1)X(2)Y(3)Y(4), \quad 5 \equiv X(1)X(2)Z(3)Y(4), \quad 6 \equiv X(1)Z(2)$
$X(3)Y(4), 7 \equiv X(1)X(2)Y(3)Z(4), S_1 \equiv \langle X|Z\rangle$, and $S_2 \equiv \langle Y|Z\rangle$. This formulation assumes that the overlap and exchange integrals between the orbitals on the atoms X and Y are negligible.

The matrix elements of the inactive electrons and the interaction between the active and inactive electrons can be approximated by expressing the corresponding potential surfaces as a quadratic expansion around the equilibrium values of the various internal coordinates, and by nonbonded potential functions for the interaction between atoms not bonded to each other or to a common atom:

$$\langle \chi_i | \mathbf{H}_{inact} | \chi_i \rangle + H'_{ii} = 1/2 \sum_m K^{(i)}_{b,m} (b - b^{(i)}_{0,m})^2$$

$$+ 1/2 \sum_m K^{(i)}_{\theta,m} (\theta - \theta^{(i)}_{0,m})^2 + U^{(i)}_{nb,inact} + U^{(i)}_{nb,inact-act}$$

$$\langle \chi_i | \mathbf{H}_{inact} | \chi_j \rangle + H'_{ij} = 0 \qquad (2.39)$$

where the b's and θ's are, respectively, the bond lengths and bond angles in both the inactive region (the R fragment) and in the intersection between the active and inactive regions (bonds and angles that connect the two regions). Similarly, the U_{nb} terms represent both the nonbonded interaction within the active region and between the active and inactive regions.

The usual EVB procedure involves diagonalizing this 3×3 Hamiltonian. However, here we wish to use a very simple model for our reaction and represent the potential surface and wavefunction of the reacting system using only two electronic states. Using a two-state system will preserve most of the important features of the potential energy surface while at the same time provide a simple model that will be more amenable to discussion than the three-state system. For the two-state system we define the following states as the reactant and product wavefunctions:

$$\psi_1 = \alpha_1(\mathbf{R})\Phi_1 + \beta_1(\mathbf{R})\Phi_3$$

$$\psi_2 = \alpha_2(\mathbf{R})\Phi_2 + \beta_2(\mathbf{R})\Phi_3 \qquad (2.40)$$

The matrix elements of the 2×2 active Hamiltonian are then given by

$$H^0_{11} = \langle \psi_1 | \mathbf{H} | \psi_1 \rangle = \alpha_1^2 \langle \Phi_1 | \mathbf{H} | \Phi_1 \rangle + \beta_1^2 \langle \Phi_2 | \mathbf{H} | \Phi_2 \rangle$$
$$+ 2\alpha_1\beta_1 \langle \Phi_1 | \mathbf{H} | \Phi_2 \rangle$$

$$H^0_{22} = \langle \psi_2 | \mathbf{H} | \psi_2 \rangle = \alpha_2^2 \langle \Phi_3 | \mathbf{H} | \Phi_3 \rangle + \beta_2^2 \langle \Phi_2 | \mathbf{H} | \Phi_2 \rangle$$
$$+ 2\alpha_2\beta_2 \langle \Phi_3 | \mathbf{H} | \Phi_2 \rangle$$

$$H^0_{12} = \langle \psi_1 | \mathbf{H} | \psi_2 \rangle = \alpha_1\alpha_2 \langle \Phi_1 | \mathbf{H} | \Phi_3 \rangle + \alpha_1\beta_2 \langle \Phi_1 | \mathbf{H} | \Phi_2 \rangle$$
$$+ \alpha_2\beta_1 \langle \Phi_2 | \mathbf{H} | \Phi_3 \rangle + \beta_1\beta_2 \langle \Phi_2 | \mathbf{H} | \Phi_2 \rangle \qquad (2.41)$$

These matrix elements are in a form that can be evaluated using standard quantum chemical methods. This evaluation is tedious and the earlier assumptions that we made will lead to significant errors in the matrix elements. On the other hand, we can conveniently use experimental information to approximate the diagonal matrix elements.

2.6. SOME RELEVANT COMPUTER PROGRAMS

2.A. A Simple LD Program

```
c        This program performs simple Langevin Dipoles(LD) calculations
c        and provides an estimate of solvation free energies.

         implicit real*8 (a-h,o-z)
         dimension iact(150),qt(150),gs(20),x(150,3)
         common/ind/xpol(3000,3),fpol(3000,3),vq(30),x0(3),npol
         x(1,1)=0.0
         x(1,2)=0.0
         x(1,3)=0.0
         qt(1)=-1
         iact(1)=4
         natom=1
         nat_all=1
         call solvate(qt,x,nat_all,iact,gs,natom)
         end
c=============================================================
         subroutine solvate(qq,x,nq,iact,et,natom)

c        This subroutine evaluates solvation energies using
c        the fixed center langevin dipoles (FCLD) method
c        fpol is the solvent dipole

c        vq       The potential at the sites of the solute atoms
c        dt       The grid spacing
c        rg       The radius of the system
         implicit real*8 (a-h,o-z)
         common/par/d(12),dx(12),rz(12),am(12),dl(5),avd(12),bvd(12),cvd(12)
         dimension qq(150),x(150,3),xp(3),rl(3),ep(3),em(3),iact(150),ru(10)
         common/ind/xpol(3000,3),fpol(3000,3),vq(30),x0(3),npol
         data ru/1.4,2.2,2.7,2.4,2.2,2.2,2.2,2.2,2.2,2.2/
         do i=1,3
             x0(i)=0.0
             do j=1,natom
                 x0(i)=x0(i)+x(j,i)
             enddo
             x0(i)=x0(i)/float(natom)
         enddo
         dt=2.7
         rg=7.
         fnl=rg/dt
         fnl=2*fnl
         nl=fnl
         nl2=nl/2+1
         et=0.
         fm0=0.35
         fv=(dt/2.7)**3
         lpol=0
```

```
      do 1 l1=1,nl                                    ! Builds the grid.
        do 1 l2=1,nl
          do 1 l3=1,nl
            xp(1)=dt*(l1-nl2)+x0(1)                   ! Generates the coordinates of the grid points.
            xp(2)=dt*(l2-nl2)+x0(2)
            xp(3)=dt*(l3-nl2)+x0(3)
            r0=100.
            rex=0.
            do 3 k=1,3
              ep(k)=0.
              rex=rex+(xp(k)-x0(k))**2
3           continue
            rg2=rg**2
            if(rex.gt.rg2) go to 1                    ! Checks if the grid points are outside the
            do 4 i=1,nq                               ! system boundary.
              r2=0.
              do 5 k=1,3                              ! Calculates the distance between the grid
                rl(k)=xp(k)-x(i,k)                    ! points and the solute.
                r2=r2+rl(k)**2
5             continue
              r1=sqrt(r2)                             ! Finds the smallest distance between the solute
              if(r1.lt.r0) r0=r1                      ! and the grid points. Checks if the given grid
              if(r0.lt.ru(iact(i))) go to 1           ! point is within the van der waals distance from
              do 6 k=1,3                              ! any of the solute atoms.
                ep(k)=ep(k)+qq(i)*rl(k)/(r2*r1)       ! Evaluates the field from the solute.
6             continue
4           continue
            epl2=0.
            do 9 k=1,3
9           epl2=epl2+ep(k)**2
            epl=sqrt(epl2)+0.0001
            xl=332*fm0*epl/(0.6*(1.+r0))+0.0001       ! Places the Langevin dipoles at grid points.
            y1=exp(xl)                                ! The (1+r0) term is a distance dependent
            y2=1/y1                                   ! dielectric constant for the non iterative LD
            cote=(y1+y2)/(y1-y2)                      ! procedure ( See references 4).
            cote=cote-1/xl
            do 8 k=1,3
              em(k)=fm0*fv*cote*ep(k)/epl
              xpol(lpol+1,k)=xp(k)                    ! Stores the langevin dipoles.
              fpol(lpol+1,k)=em(k)
8           continue
            lpol=lpol+1
            if(lpol.gt.3000) print 1003,lpol
            do 7 k=1,3                                ! Calculates dipole-field interaction and
7           et=et-166*em(k)*ep(k)                     ! the corresponding energy spent on
1     continue                                        ! polarizing the solvent.
      print 1000,et
      do 100 j=1,nq                                   ! Calculates the potential from the solvent
        vq(j)=0.                                      ! dipoles at the sites of the solute atoms.
        do 100 i=1,lpol
          r2=0.
          rm=0.
          do 101 k=1,3
```

```
                    rl(k)=xpol(i,k)-x(j,k)
                    rm=rm+rl(k)*fpol(i,k)
101                 r2=r2+rl(k)**2
                    rx=sqrt(r2)
100     vq(j)=vq(j)+332*rm/rx**3
        print 1004 ,(vq(j),j=1,nq)
1000    format(2x,'estimate of solvation energy',f10.2,/)
1003    format(2x,'lpol too large')
1004    format(2x,'potential on solutes atoms (kcal/mol)',5f10.2)/
        return
        end
```

2.B. An EVB Program

```
c       This program uses THE LANGEVIN DIPOLE MODEL to evaluate
c       the EVB potential surfaces for reactions in solutions.

c       dx      The dissociation energy of the mixed Morse potential
c       d       The dissociation energy of the pure covalent Morse potential
c       dl      The gas phase energy of the given resonance structure where the fragments are at infinite seperation.
c       gs      The solvation energy
c       h(i,i)  The hamiltonian matrix elements
c       ibt     Defines the bonding in each resonance structure.
c       ia      Defines the atom type
c       natom   The number of atoms
c       nb      The number of bonds
c       nr      The number of resonance structures
c       q       The atomic charges
c       rz      The bond length
c       u       The eigenvectors of the h matrix
c       x       The coordinate vector (x,y,z for each atom)
        implicit real*8(a-h,o-z)
        common/h1/h(20,20),e(20),egas(20)
        common/topol/ib(20,20),jb(20,20),ia(20,20),q(20,150),x(150,3),
     &          lb(20,2),ibt(20,20),ipt(20,2),ltt,at(10),att(150)
        common/par/d(12),dx(12),rz(12),am(12),d1(5),avd(12),bvd(12),cvd(12)
        dimension u(20,20),ex(20),p(20),ev(20),gs(20),qg(150)
        character at1*3,at2*3,at*3,att*3,at11*3,at22*3
        data at/'H','O','N','C','S','Cl','Na','F','Br','I'/
        ltt=0
1       read(5,*) at1,at2
        print '(2x,a,2x,a)',at11,at22
        itt=iat(at1)
        if(at1.eq.'*') go to 2
        write(6,*) 'iat',itt
        ltt=ltt+1
        ipt(ltt,1)=iat(at1)
        ipt(ltt,2)=iat(at2)
        read (5,*) dx(ltt),d(ltt),am(ltt),rz(ltt),avd(ltt),bvd(ltt),cvd(ltt)
        go to 1
2       continue
        read(5,*) natom,nr
```

```
        nb=0
        do i=1,natom-1                              ! Finds all possible pairs of bonded atoms.
            do j=i+1,natom
                nb=nb+1
                lb(nb,1)=i
                lb(nb,2)=j
            enddo
        enddo
        write(6,*) 'lb',(lb(l,1),lb(l,2),l=1,nb)
        do ii=1,nr                                  ! Defines the bonding pattern
            read(5,*) nbt,(ib(i,ii),jb(i,ii),i=1,nbt)   ! in each resonance structure.
            do 3 l=1,nb
                ibt(l,ii)=0
                do 3 i=1,nbt
                    if(ib(i,ii).eq.lb(l,1).and.jb(i,ii).eq.lb(l,2)) then
                        ibt(l,ii)=1
                    endif
                    if(ib(i,ii).eq.lb(l,2).and.jb(i,ii).eq.lb(l,1)) then
                        ibt(l,ii)=1
                    endif
3               continue
            write(6,*) 'ibt' ,(ibt(l,ii),l=1,nb)
        enddo
        do 4 i=1,nr
            read (5,*) (att(k),k=1,natom)
            do k=1,natom
                ia(i,k)=iat(att(k))
            enddo
4       continue
        do 5 i=1,nr
            read (5,*)(q(i,l),l=1,natom)
5       continue
        read(5,*) (d1(ii),ii=1,nr)
        read(5,*) ((x(i,k),k=1,3),i=1,natom)
        read(5,*) n_ext                             ! Read positions and charges of external atoms.
        do i=1,n_ext                                ! This option is used for calculations of reaction in
            read(5,*) (x(i+natom,k),k=1,3), q(1,i+natom)   ! protein.
        enddo
        do 10 ii=1,nr                               ! Evaluates  the diagonal energies of  the
            e(ii)=d1(ii)                            ! gas phase Hamiltonian.
            ex(ii)=0.
            do 11 k=1,nb
                ib1=ib(k,ii)
                jb1=jb(k,ii)
                if(ib1.eq.0) go to 11               ! Evaluates the bond energies.
                ia1=ia(ii,ib1)
                ia2=ia(ii,jb1)
                call rr(x,ib1,jb1,r)
                fmt=fm(r,ia1,ia2,1)
                fmtx=fm(r,ia1,ia2,2)
                ex(ii)=ex(ii)+fmt
11          continue
            call gqq(ii,ii,natom,eq,evd,nb)         ! Evaluates the nonbonded energies .
            e(ii)=e(ii)+eq+evd
            gs(ii)=0.
```

```
10        egas(ii)=e(ii)+ex(ii)
          nxs=2                                                    ! Evaluates the LD solvation energy
          nloop1=5                                                 ! of each resonance structure.
          do 40 iloop=1,nloop1                                     ! The initial polarization is determined
              call gsol(nxs,iloop,gs,natom,n_ext,nr,q,qg,x)        ! by the charges of the nxs states.
              do 15 ii=1,nr
                  print1111,gs(ii)
15            h(ii,ii)=egas(ii)+gs(ii)
              call offdia(nr,nb)                                   ! Evaluates the off-diagonal terms.
              print 1041,(egas(i),i=1,nr)
              do 20 i = 1,nr
                  print 1004,(h(i,j),j=1,nr)
20            continue
              call diag(h,u,ev,nr)
              do 33 id=1,nr
                  p(id)=u(id,nr)*u(id,nr)
33            continue
              print 1042,(ev(id),id=1,nr)
              print 1143,(p(id),id=1,nr)
              nxso=nxs
              vt1=0.
              do 90 i=1,nr
                  vt=abs(u(i,nr))
                  if(vt.le.vt1) go to 90
                  nxs=i                                            ! Determines which eigenvector corresponds
                  vt1=vt.                                          ! to the current ground state.
90            continue
              do 91 k=1,natom                                      ! Evaluates the ground state charges.
                  qg(k)=0.
                  do 91 i=1,nr
                      qg(k)=qg(k)+q(i,k)*u(i,nr)**2
91            continue
40        continue
1004      format(1x,10f10.2)
1040      format(2x,'resonance structures included ',/,10i10)
1041      format(2x,'gas phase diagonal energies',/,10f8.1,/)
1042      format(2x,'eigen values',/,10f8.1)
1143      format(' ground state wave function',/,10f8.4)
1044      format(2x,'the gas phse deltas',/,10f8.2)
1111      format(2x,'gs=',f10.4)
          end
c========================================================
          subroutine offdia(nr,nb)

c         Evaluates the off diagonal matrix elements
          implicit real*8 (a-h,o-z)
          common/h1/h(20,20),e(20),egas(20)
          common/topol/ib(20,20),jb(20,20),ia(20,20),q(20,150),x(150,3),
     &             lb(20,2),ibt(20,20),ipt(20,2),ltt,at(10),att(150)
          common/par/d(12),dx(12),rz(12),am(12),d1(5),avd(12),bvd(12),cvd(12)
          do 30 ii=1,nr                                            ! Identifies pairs of resonance structures which
              ii1=ii+1                                             ! can be converted to each other by breaking
              do 30 jj=ii1,nr                                      ! a single covalent bond and forming an ion pair.
                  if(jj.gt.nr) go to 30                            ! such resonance structures are coupled to each
                  h(ii,jj)=0.                                      ! other by eq. (1.59) ( see ref. 6 for details) .
```

```
                    ix=0
                    do 31 k=1,nb
                        it1=ibt(k,ii)
                        it2=ibt(k,jj)
                        is=it1-it2
                        if(is.eq.0) go to 31
                        ix=ix+1
                        if(is.eq.1) ij=jj
                        if(is.eq.(-1)) ij=ii
                        ik=ii
                        if(is.eq.(-1)) ik=jj
                        kt=k
31                  continue
                    i1=lb(kt,1)
                    j1=lb(kt,2)
                    if(ix.ne.1) go to 33
                    cq1=q(ij,i1)-q(ik,i1)
                    if(cq1.eq.0.0) go to 33
                    cq2=q(ij,j1)-q(ik,j1)
                    if(cq2.eq.0.0) go to 33
                    cq=cq1*cq2
                    qq=q(ij,i1)*q(ij,j1)          ! Finds the difference in the charge of a given
                    ia1=ia(ij,i1)                 ! atom in the two resonance structures.
                    ia2=ia(ij,j1)
                    call rr(x,i1,j1,r)
                    e1=egas(ik)
                    e2=egas(ij)
                    del=d1(ij)-d1(ik)
                    idpf=1
                    h(ii,jj)=fl(r,del,ia1,ia2,idpf,,e1,e2,qq)
33                  h(jj,ii)=h(ii,jj)
30          continue
            return
            end
c=====================================================
            function fnonb(r,ia1,ia2,idpf)

            implicit real*8 (a-h,o-z)
            common/par/d(12),dx(12),rz(12),am(12),d1(5),avd(12),bvd(12),cvd(12)
            ic=icode(ia1,ia2)
            A1=avd(ic)
            b=bvd(ic)
            cc=cvd(ic)
            fnonb=A1*exp(-b*r)+cc*r**(-9)
            return
            end
c=====================================================
            function fm(r,ia1,ia2,ig)

            implicit real*8 (a-h,o-z)
            common/par/d(12),dx(12),rz(12),am(12),d1(5),avd(12),bvd(12),cvd(12)
            ic=icode(ia1,ia2)
            dmn=d(ic)
```

```
          if(ig.eq.2) dmn=dx(ic)
          fa=am(ic)
          r0mn=rz(ic)
          dr=r-r0mn
          ex1=exp(-fa*dr)
          ex2=ex1*ex1
          fm=dmn*(ex2-2*ex1)
          return
          end
c=================================================
          subroutine rr(x,i,j,r)

          implicit real*8 (a-h,o-z)
          dimension x(60,3)
          r2=0.
          do 1 k =1,3
               r2=r2+(x(j,k)-x(i,k))**2
1         continue
          r=sqrt(r2)
          return
          end
c=================================================
          function fl(r,del,ia1,ia2,idpf,,e1,e2,qq)

          implicit real*8 (a-h,o-z)
c         off diagonal bond coupling
          fmx=fm(r,ia1,ia2,2)
          fm1=fm(r,ia1,ia2,1)
          d1=fm1-fmx
          fnon=fnonb(r,ia1,ia2,idpf)
          d2=del+qq*332/r+fnon-fmx
          f=d1*d2
          f=abs(f)
          fl=sqrt(f)
          return
          end
c=================================================
          subroutine gsol(iis,iloop,gs,natom,n_ext,nr,q,qg,x)

c         this subroutine controls the calculations of the solvation energies.

          implicit real*8 (a-h,o-z)
          common/topol/ib(20,20),jb(20,20),ia(20,20),qd(20,150),xd(150,3),
     &              lb(20,2),ibt(20,20),ipt(20,2),ltt,at(10),att(150)
          dimension iact(150),qt(150),gs(20),q(20,150),x(150,3)
          dimension qg(150)
          common/ind/xpol(3000,3),fpol(3000,3),vq(150),x0(3),npol
          do 1 i=1,natom
               iact(i)=ia(iis,i)
               qt(i)=qg(i)
               if(iloop.eq.1) qt(i)=q(iis,i)
1         continue
          do i= 1,n_ext
```

```
        qt(i+natom)=q(1,i+natom)
        iact(i+natom)=2
     enddo
     nat_all=natom+n_ext
     call solvate(qt,x,nat_all,iact,et,natom)          ! Evaluates the polarization of the solvent
     gext=0.0                                           ! dipoles in the presence of the ground state
     do i=1,n_ext                                       ! charges of the solute.
        gext=gext+vq(i+natom)*qt(i+natom)
     enddo
     do 2 it=1,nr
        gs(it)=-et+gext
        do 2 i=1,natom
           gs(it)=gs(it)+vq(i)*q(it,i)
2    continue
     do 3 it=1,nr                                       ! Evaluates the interactions between the solute
        do j=1,n_ext                                    ! and the external charges.
           do i=1,natom
              r2=0.
              do k=1,3
                 r2=r2+(x(i,k)-x(j+natom,k))**2
              enddo
              r1=sqrt(r2)
              gs(it)=gs(it)+332*q(it,i)*q(1,j+natom)/r1
              if(it.eq.1) vq(i)=vq(i)+332*q(1,j+natom)/r1
           enddo
        enddo
3    continue
     return
     end
```

c===

```
     subroutine solvate
     The corresponding subroutine is given in 2.A
```

c===

```
     subroutine gqq(ii,id,natom,eq,evd,nb)

     implicit real*8 (a-h,o-z)
     common/topol/ib(20,20),jb(20,20),ia(20,20),q(20,150),x(150,3),
    &            lb(20,2),ibt(20,20),ipt(20,2),ltt,at(10),att(150)
c    evalutes the charge-cherge interactions between
c    the nonbonded atoms in each resonance structure.
     eq=0.
     evd=0.
     do 1 iq=1,natom
        i1=iq+1
        do 1 jq=i1,natom
           if(i1.gt.natom) go to 1
           do 10 kb=1,nb
              if(ib(kb,ii).eq.jq.and.jb(kb,ii).eq.iq) go to 1   ! Checks whether the given pair of atoms are
              if(ib(kb,ii).eq.iq.and.jb(kb,ii).eq.jq) go to 1   ! bonded to each other.
10         continue
           r2=0.
           do 4 k=1,3                                           ! Calculates the distance between nonbonded atoms.
```

```
4                    r2=r2+(x(iq,k)-x(jq,k))**2
                     r=sqrt(r2)
                     eqt=eq
                     eq=eq+332.*q(id,iq)*q(id,jq)/r
                     eqt=eq-eqt
                     ia1=ia(ii,iq)
                     ia2=ia(ii,jq)
                     idpf=1
                     fvd=fnonb(r,ia1,ia2,idpf)
                     evd=evd+fvd
1        continue
         return
         end
c=================================================
         function iat(att)

         implicit real*8 (a-h,o-z)
         common/topol/ib(20,20),jb(20,20),ia(20,20),q(20,150),x(150,3),
     &                lb(20,2),ibt(20,20),ipt(20,2),ltt,at(10),att(150)
         character at*3,att*3
         do i =1,10                                   ! Determines the atom type.
             if(att.eq.at(i)) iat=i
         enddo
         return
         end
c=================================================
         function icode(i1,j1)
         implicit real*8 (a-h,o-z)
         common/topol/ib(20,20),jb(20,20),ia(20,20),q(20,150),x(150,3),
     &                lb(20,2),ibt(20,20),ipt(20,2),ltt,at(10),att(150)
         character at*3
         do i=1,ltt
             if(ipt(i,1).eq.i1.and.ipt(i,2).eq.j1) icode=i
             if(ipt(i,1).eq.j1.and.ipt(i,2).eq.i1) icode=i
         enddo
         return
         end
c=================================================
         subroutine Diag
         This subroutine is given in 1.A
c=================================================
c=================================================
```

Data for (2.B) the EVB program

```
'C' 'O'
92. 69. 1.87 1.43 5200. 2.5 100.
'*' '*'
2 2     natom nr
1 1 2   nbt  ib jb
0 0 0   nbt  ib jb
'C' 'O'
'C' 'O'
0.1 -0.1
1.0 -1.0
```

```
0.0  139
0.0  0.0   0.0  coordinates (x,y,z)
1.43 0.0   0.0
0
0
```

2.C. MO Calculations Combined with the LD Solvent Model

Use here the main program from B.1. The only modification needed is to
insert before calling DIAG the following
```
if (isol.eq.1) call solvate (q,x,n,l,et)
do  i1=1,n
      h(i1,i1)=h(i1,i1)-vq(i1)/23.06
enddo
```

c==

subroutine huckel_mat(h,r,q,n,a)
The corresponding subroutine is given in 1.B

c==

subroutine scf_mat(h,r,q,n,a)
The corresponding subroutine is given in 1.B

c==

subroutine pmat(c,p1,q,nn,n,aa)
The corresponding subroutine is given in 1.B

c==

subroutine diag(h,c,ev,n)
The corresponding subroutine is given in 1.A

c==

subroutine solvate (qq,x,nq,iact,et,code)
The corresponding subroutine is given in 2.A

c==
c==

Data for (2.C) MO(scf) calculations in LD solvent model

```
'scf'
2     natom
0.0   a
1.0   cc
0.01  aa
0.0   0.0   0.0   coodinates (x,y,z for each atom)
1.43  0.0   0.0
-10.67  -15.85   alf0
1.  1.   z
1.0   0.0      the initial bond order matrix
0.0   1.0
-7.5  1.5   0.6   beta  d  mu
0  2    electronic occupation
1      isol
```

Data for (2.C) MO(huckel) calculations in LD solvent model

```
'huckel'
2    natom
1.0  a
0.0  cc
0.1  aa
0.00   0.0   0.0      coordinates
1.43   0.0   0.0
-11.25  -13.6    alf0
1.  1.   z
1.0   0.0     initial bond order matrix
0.0   1.0
-11.5   1.6   0.5    beta  d  mu
0   2     electronic occupation
1     isol
```

REFERENCES

1. (a) S. Glasstone, K. J. Laidler, and H. Eyring, *The Theory of Rate Processes*, McGraw-Hill, New York, 1941. (b) H. Eyring, J. Walter, and G. E. Kimball, *Quantum Chemistry*, Wiley, New York, 1967. (c) H. S. Johnston, *Gas Phase Reaction Rate Theory*, Ronald Press, New York, 1966.

2. (a) J. B. Anderson, *J. Chem. Phys.* **58**, 4684 (1973). (b) D. Chandler, *J. Chem. Phys.*, **68**, 2959 (1978). (c) C. H. Bennet in *Algorithms for Chemical Computations*, R. E. Christofferson, Ed., ACS, Washington, D.C., 1977 pp 63. (d) J-K. Hwang, S. Creighton, G. King, D. Whitney, and A. Warshel, *J. Chem. Phys.*, **89**, 859 (1988). (e) E. K. Grimmelman, J. C. Tully, and E. Helfand, *J. Chem. Phys.* **74**, 5300 (1981).

3. L. Onsager, *J. Am. Chem. Soc.*, **58**, 1486 (1936).

4. A. Warshel and S. T. Russell, *Quart. Rev. Biophys.*, **17**, 283 (1984).

5. C. A. Coulson and U. Danielsson, *Ark. Fys.*, **8**, 239 (1954).

6. A. Warshel and R. M. Weiss, *J. Am. Chem. Soc.*, **102**, 6218 (1980).

7. J.-K. Hwang, G. King, S. Creighton, and A. Warshel, *J. Am. Chem. Soc.*, **110**, 5297 (1988).

8. O. Tapia and D. Goscinski, *Mol. Phys.*, **29**, 1653 (1975).

$$3$$

CHEMICAL REACTION IN ALL-ATOM SOLVENT MODELS

3.1. ALL-ATOM SOLVENT MODELS

3.1.1. Explicit Models for Water Molecules

In the previous chapter we considered a rather simple solvent model, treating each solvent molecule as a Langevin-type dipole. Although this model represents the key solvent effects, it is important to examine more realistic models that include explicitly all the solvent atoms. In principle, we should adopt a model where both the solvent and the solute atoms are treated quantum mechanically. Such a model, however, is entirely impractical for studying large molecules in solution. Furthermore, we are interested here in the effect of the solvent on the solute potential surface and not in quantum mechanical effects of the pure solvent. Fortunately, the contributions to the Born–Oppenheimer potential surface that describe the solvent–solvent and solute–solvent interactions can be approximated by some type of analytical potential functions (rather than by the actual solution of the Schrodinger equation for the entire solute–solvent system). For example, the simplest way to describe the potential surface of a collection of water molecules is to represent it as a sum of two-body interactions (the interactions between pairs of atoms) plus three-body corrections (the change in the interactions between pairs of atoms due to the presence of other atoms). That is, one can use the potential surface

$$U = \sum_{i,j} (U_{nb}^{ij} + U_{qq}^{ij}) + U_{ind} \tag{3.1}$$

where the two-body terms U^{ij} are given as the sum of the contributions (in kcal/mol)

$$U_{nb}^{ij} = \sum_{k(i)k'(j)} (A_k A_{k'}/r_{kk'}^{12} - B_k B_{k'}/r_{kk'}^{6})$$

$$U_{qq}^{ij} = \sum_{k(i)k'(j)} 332 q_k q_{k'}/r_{kk'} \tag{3.2}$$

Here k runs over the atoms of the ith water molecule and the intramolecular potential of each water molecule is kept constant (for simplicity we treat the water molecules as rigid bodies). The q's are the residual charges on the hydrogen and oxygen atoms (taken, as a first approximation, to reproduce the observed gas-phase dipole moment of a water molecule), and the A and the B terms represent, respectively, the hard core repulsion and van der Waals attractions between the indicated atoms. The distances are given in Å and charges in au. The potential U_{ind} simulates the three-body inductive effect associated with the polarization of the electrons on each atom in the presence of the residual charges and induced dipoles of the remaining system. This complicated polarization effect can be simulated quite simply by the self-consistent iterative equation (see Ref. 1 for detailed derivation)

$$\mu_i^n = \gamma_i \xi_i^n$$

$$U_{ind}^n = -166 \sum_i \mu_i^n \xi_i^0$$

$$\xi_i^n = \sum_{j \neq i} q_j \mathbf{r}_{ji}/r_{ji}^3 - \sum_{j \neq i} [\mu_j^{n-1} - 3(\mathbf{r}_{ji}\mu_j^{n-1})\mathbf{r}_{ji}/r_{ij}^2]/r_{ji}^3 \tag{3.3}$$

where $\mathbf{r}_{ij} = \mathbf{r}_j - \mathbf{r}_i$, the γ's are the atomic polarizabilities, the μ_i's are the molecular induced dipoles, and ξ_i is the field on the oxygen of the ith water molecule from all the other water molecules. The superscript n indicates that we are dealing with an iterative self-consistent procedure, starting the first iteration with the field, ξ^0, due only to the permanent charge distribution (the q's), and then including the field from the induced dipoles. The units in eq. (3.3) are kcal/mol, Å and au for energy, distance, and charge, respectively.

TABLE 3.1. Parameter Sets for potential surfaces of liquid water[a]

Set 1	q	A	B	γ
Oxygen	−0.82	861.0	26.0	0
Hydrogen	0.42	0.419	0.575	0
Set 2	q	A	B	γ
Oxygen	−0.6668	830.0	25.6	1.10
Hydrogen	0.3334	0.435	0.585	0.20

[a]Charges are given in electron charge units, polarizabilities in Å^3, while A and B are given, respectively in kcal $\text{mol}^{-1}\,\text{Å}^{12}$ and kcal $\text{mol}^{-1}\,\text{Å}^{6}$. Each water molecule is treated as a rigid body with O–H bond length of 1.0 Å and H–O–H angle of 109.5°.

3.1.2. How to Obtain Refined Potential Surfaces for the Solvent Molecules

The potential parameters (q, A, B, and γ) can be obtained by using eq. (3.1) to evaluate the potential surface of two interacting water molecules, and then adjusting these parameters to obtain the best fit between the calculated surface and the corresponding surface evaluated by performing accurate quantum mechanical calculations (e.g., Ref. 2). Further refinement of the parameters is then obtained by fitting calculated and observed properties of bulk water. This fitting procedure may be augmented by using the same model to calculate the properties of the various forms of the ice phase (e.g., Ref. 3). Typical sets of refined parameters are given in Table 3.1.

The parameter sets of Table 3.1 and the potential functions of eq. (3.1) provide an approximate description of the potential surface associated with a collection of water molecules.

3.2. EXPLORING THE SOLVENT PHASE SPACE BY THE METHOD OF MOLECULAR DYNAMICS

3.2.1. Statistical Mechanics and the Relationship Between Macroscopic and Microscopic Properties

With a given set of potential functions we can evaluate various average properties of the solvent. In particular, we would like to simulate experimentally observed macroscopic properties using microscopic solvent models. To do this we have to exploit the theory of statistical mechanics which tells us that the average of a given property, A, which is independent of the momenta of the system, is given by (Ref. 4):

$$\langle \mathbf{A} \rangle = \int \mathbf{A}(r)P(\mathbf{r})\,d\mathbf{r} = \int \mathbf{A}(\mathbf{r})\exp\{-U(\mathbf{r})\beta\}\,d\mathbf{r}/z(U) \qquad (3.4a)$$

$$z(U) = \int \exp\{-U(\mathbf{r})\beta\}\,d\mathbf{r} \qquad (3.4b)$$

where $d\mathbf{r}$ designates the volume element of the complete space spanned by the $3n$ vector $\mathbf{r}$ associated with the n atoms of the system. The evaluation of eq. (3.4) requires us to explore all the points in the entire configuration space of the system. Such a study is clearly impossible with any of the available computers. However, we can hope that the average over a limited number of configurations will give similar results to those obtained from an average over the entire space. With this working hypothesis we can try to look for an efficient way of spanning phase space. With present-day computers, we can span a significant number of configurations by the so-called Monte Carlo (MC) (Ref. 5) methods or the methods of molecular dynamics (MD) (Ref. 6) outlined below.

3.2.2. Molecular Dynamics and Simulations of Average Solvent Properties

In MD simulations we simply solve numerically the classical equations of motion, expressing the changes in coordinates and velocities at a time increment Δt by

$$r_i(t + \Delta t) = r_i(t) + \dot{r}_i(t)\,\Delta t$$

$$\dot{r}_i(t + \Delta t) = \dot{r}_i(t) + \ddot{r}_i(t)\,\Delta t = \dot{r}_i(t) - [(\partial U/\partial r_i)m_i^{-1}]\,dt \qquad (3.5)$$

where the dot designates a time derivative and we use Newton's law:

$$m\ddot{r}_i = F_i = -\partial U/\partial r_i \qquad (3.6)$$

Starting with a given set of initial conditions [e.g. with the values of $r_i(t = 0)$ and $\dot{r}_i(t = 0)$] we can evaluate $\mathbf{r}(t)$ either by numerically integrating eq. (3.5) or by using the somewhat more complicated but far better approximation (Ref. 6)

$$r_i(t + \Delta t) = r_i(t) + \dot{r}_i\,\Delta t + [4\ddot{r}_i(t) - \ddot{r}_i(t - \Delta t)]\,\Delta t^2/6$$

$$\dot{r}_i(t + \Delta t) = \dot{r}_i + [2\ddot{r}(t + \Delta t) + 5\ddot{r}_i(t) - \ddot{r}_i(t - \Delta t)]\,\Delta t/6 \qquad (3.7)$$

This equation allows one to obtain much more accurate results than those of eq. (3.5), using the same Δt's.

The time-dependent coordinate vector $\mathbf{r}(t)$ describes the position of each atom at the time t, and is known as a *classical trajectory*.

Exercise 3.1.   Consider a one-dimensional harmonic oscillator with $U = 0.5$ $(X - 2)^2$, $m = 1$, $X(0) = 1$, $\dot{X}(0) = 0$ and evaluate X and $\dot{X}$ for $t = 0, 0.1$, and 0.2 using eq. (3.5).

Solution 3.1.   The first three steps in the integration of eq. (3.5) are summarized in Table 3.2. Follow the example of the table for a few more steps.

The propagation of classical trajectories of the atoms of a given system corresponds to a fixed total energy (determined by the specified initial conditions). However, the evaluation of statistical mechanical averages (e.g., eq. 3.4) implies that the system included in the simulation is a part of a much larger system (ensemble) whose atoms are not considered in an explicit way. Thus, in order to simulate a given macroscopic property at a specified temperature we must introduce some type of a "thermostat" in the system that will keep it at the given temperature. This can be easily accomplished by assuming equal partition of kinetic energy among all degrees of freedom. Since each atom has three degrees of freedom with kinetic energy of $\frac{1}{2}m\dot{r}^2 = \frac{3}{2}k_B T$ we obtain:

$$T = \sum_i m_i \dot{r}_i^2 / (3nk_B) \tag{3.8}$$

where n is the number of atoms in our system. In general, we can adjust the temperature during the simulation by scaling the velocities. That is, when T is smaller than the target temperature we can scale $\dot{r}$ uniformly by $(1 + \varepsilon)$ until the target temperature is obtained. If T is higher than the target temperature, then a scaling of $(1 - \varepsilon)$ is used.

Exercise 3.2.   Write a computer program that solves Exercise 3.1 for any given time and evaluate the kinetic energy and potential energy as a function of time.

TABLE 3.2.   The First Three Steps in Exercise 3.1[a]

t	X	$\dot{X}$
0	1	0
0.1	1	0.1
0.2	1.01	0.2
0.3	1.03	0.299

[a]Please continue for a few more steps (if you needed this help).

With a computer program that evaluates $\mathbf{r}$ as a function of time for a given $U(\mathbf{r})$ we can use the *ergodic hypothesis* (which states that the time average over a long time is equal to the configuration average) and write

$$\langle A \rangle = \sum_i^N A(t_i)/N \tag{3.9}$$

The desired average is simply obtained by a time average of the given property. For example, one of the interesting properties of bulk solvents is the radial distribution function (rdf), which expresses the probability of finding a given atom type around a reference atom by

$$g(r)_{AB} = \langle N_B(r, r + \Delta r) \rangle /(4\pi\rho_B r^2 \Delta r) \tag{3.10}$$

where A and B refers to atom type, $N(r, r + \Delta r)$ is the average number of atoms of the B type (e.g., an oxygen atom) found at a distance between r to $r + \Delta r$ from the reference atom, and ρ_B is the bulk number density of the Bth type atom. Such an average for a water model that uses the parameters of Table 3.1 is compared in Fig. 3.1 to the corresponding experimental $g(r)$ (for early related studies see Ref. 7). Obviously, one can refine the

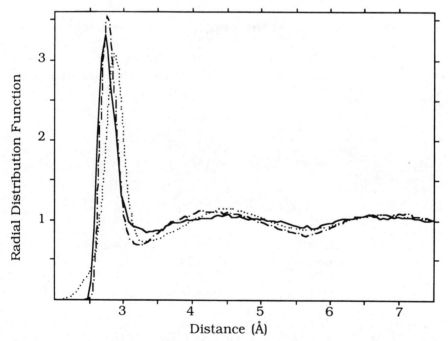

FIGURE 3.1. The oxygen–oxygen radial distribution function of water. The dotted curve represents the experimental result and the other curves correspond to the results calculated with the different models considered in Ref. 10.

parameters in the solvent potential functions by improving the agreement between the calculated and observed $g(r)$ as well as other properties.

Exercise 3.3. Write a computer program for N dipoles (represented by pairs of bonded atoms 0.9 Å apart with charges of $+0.4$ and -0.4 au). Use the potential

$$U = \sum_{ij} 332(q_iq_j/r_{ij}) + U_{nb}(r_{ij}) + \sum_{ij}^{\text{dipoles}} 400(r_{ij} - 2)^2$$

$$U_{nb}(r) = 40.000r^{-12} - 288r^{-6} \qquad (3.11)$$

where the first sum involves only atoms in different dipoles, while the second involves only the bonded atoms in each dipole. Run MD simulations with this potential and evaluate the radial distribution function for this system.

Solution 3.3. Use Program 3.A.

3.3. CALCULATION OF SOLVATION ENERGIES BY FREE-ENERGY PERTURBATION METHODS

3.3.1. Direct Calculations of Free Energy Converge Very Slowly

With a realistic solvent model, we can explore the properties of solvated molecules. As before, we take a classical approach by adding the solute–solvent interaction term (U_{Ss}) to the potential surface of the system and write

$$U = U_{SS} + U_{Ss} + U_{ss} + U_{ind} \qquad (3.12)$$

where S and s designate solute and solvent, respectively, and U_{Ss} is represented by

$$U_{Ss} = \sum_{i(S)j(s)} (U_{nb}^{ij} + U_{Qq}^{ij}) \qquad (3.13)$$

here the nonbonded potential U_{nb} is represented by the same function as in eq. (3.2) and the term U_{Qq} is given by

$$U_{Qq}^{ij} = 332Q_iq_j/r_{ij} \qquad (3.14)$$

The term U_{ss} is the solvent–solvent interaction term [the U_{nb} and U_{qq} terms of eq. (3.1)] and U_{ind} is the induced dipoles three-body term which includes now the field both from the solute and the solvent. With a potential surface for a solvated solute we can address the important issue of evaluating solvation energies. In principle, one can try to evaluate the average potential energy of the solvated molecules relative to its energy in the gas phase by

$$\Delta U_{\text{solvation}} = \langle U_{Ss} + U_{ss} + U_{\text{ind}} + U_{SS} \rangle_{\text{sol}} - \langle U_{SS} \rangle_{\text{gas}} \quad (3.15)$$

where $\langle \ \rangle_{\text{sol}}$ indicates that the corresponding average is evaluated in solution. The convergence of such calculations is, however, very slow.

One may also try to evaluate solvation-free energies by using the corresponding partition function through the formula (Ref. 4)

$$\Delta G_{12} \simeq \Delta A_{12} = -\beta^{-1}[\ln\{z(U_2)/z(U_1)\}] \quad (3.16)$$

where the z's of eq. (3.4b) are evaluated by a direct-phase space exploration. Such calculations, however, converge even more slowly than those based on eq. (3.15).

Note that eq. (3.16) involves the Helmholtz function, A, rather than the Gibbs function G, but the difference between ΔA and ΔG is negligible in solutions and we will use ΔG in this book.

Exercise 3.4. Evaluate the free-energy difference between two one-dimensional harmonic oscillators with potentials $U_1 = (X - 1)^2$ and $U_2 = 0.3(X - 1.2)^2$, respectively.

Solution 3.4. Evaluate the corresponding $z(U_1)$ and $z(U_2)$ with eq. (3.4b) using a numerical integration procedure and then use eq. (3.16).

3.3.2. Perturbation Calculations of Free-Energy Changes

Fortunately, calculations of solvation-free energies appear to converge faster if one uses the so-called *free-energy perturbation (FEP) method* [also related to the *umbrella sampling method* (Ref. 8)]. This method evaluates the free energy associated with the change of the potential surface from U_1 to U_2 by gradually changing the potential surface using the relationship

$$U_m(\lambda_m) = U_1(1 - \lambda_m) + U_2\lambda_m \quad (3.17)$$

The free-energy increment $\delta G(\lambda_m \to \lambda_{m'})$ associated with the change of U_m to $U_{m'}$ can be obtained by examining the corresponding change in the partition function of the system. That is,

$$z(U_{m'})/z(U_m) = \int \exp\{-U_{m'}\beta\} \, d\mathbf{r} \Big/ \int \exp\{-U_m\beta\} \, d\mathbf{r}$$

$$= \int \exp\{-(U_{m'} - U_m)\beta\} \, d\mathbf{r} \left[\exp\{-U_m\beta\} \Big/ \int \exp\{-U_m\beta\} \, d\mathbf{r} \right] \quad (3.18)$$

This result is precisely the average of $\exp\{-(U_{m'} - U_m)\beta\}$ over phase space, since the factor in square brackets is the statistical $(\exp\{-U\beta\}/z)$

factor of eq. (3.4). Running classical trajectories on the potential U_m, samples the space according to the factor in square brackets. Thus we can use the time average over these trajectories and write

$$\exp\{-\delta G(\lambda_m \to \lambda_{m'})\beta\} = (z(U_{m'})/z(U_m)) = \langle \exp\{-(U_{m'} - U_m)\beta\}\rangle_m$$
(3.19a)

where $\langle\ \rangle_m$ indicates that the given average is evaluated by propagating trajectories over U_m. The result of eq. (3.19a) can be expressed for small changes in λ_m by expanding the exponential expression as

$$\delta G(\lambda_m \to \lambda_{m'}) = \langle U_{m'} - U_m\rangle_m$$

$$(\lambda_m - \lambda_{m'}) \to 0$$
(3.19b)

This means that the free-energy change associated with small perturbations of the potential U can be evaluated by running trajectors with the initial potential and calculating the average value of the difference between the new potential and the initial potential. The overall free energy change is now obtained by changing λ_m in n equal increments and evaluating the sum of the corresponding δG's:

$$\Delta G(U_1 \to U_2) = \sum_{m=0}^{n-1} \delta G(\lambda_m \to \lambda_{m+1})$$
(3.20)

Further clarification of this procedure can be obtained by examining the free energy associated with changing the charge of a given atom in solution. To do this, we change the charges gradually using eq. (3.17), where U_1 and U_2 are the potentials associated with the neutral and charged atom, respectively. Equation (3.20) is used then to evaluate the reversible work associated with the charging processes. A typical example is given in Fig. 3.2. The result of this charging process [which is also referred to as an "*adiabatic charging*" (Ref. 10)] is not much different than that obtained by macroscopic approaches where the free energy is proportional to the square of the charge by the well-known Born's formula (Ref. 9)

$$\Delta G(Q) = -166(Q^2/\bar{a})(1 - 1/d)$$
(3.21)

where $\bar{a}$ is the radius of the cavity formed by the macroscopic solvent around the charged atom and d is the dielectric constant of the solvent [ΔG, a, and Q are given in kcal/mol, Å and au, respectively]. Note, however, that the cavity radius and the dielectric constant are not given from microscopic considerations.

Apparently the rigorous all-atom FEP approach reflects a rather simple physics; The solvent polarization responds linearly to the development of charges on the solute atoms (Ref. 1). This is why the simple LD model gives similar results to those obtained by the FEP approach (see Ref. 10).

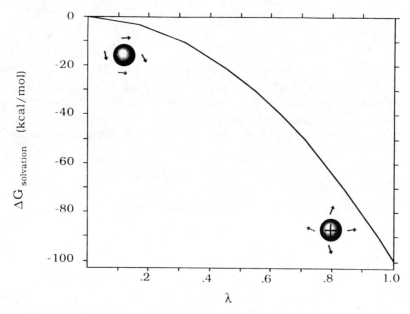

FIGURE 3.2. A free-energy perturbation calculation of the free energy associated with an adiabatic charging of an Na^+ ion in water (Ref. 10). The parameter λ transforms the solute from it uncharged ($\lambda = 0$) to its charged form ($\lambda = 1$).

Exercise 3.5. Consider the simple case of a positive ion surrounded by 10 dipoles, using for the dipole–dipole interaction the potential of Exercise 3.3 and for the ion–dipole interaction the potential

$$U_{Ss} = \sum_i (332Q_1 q_i / r_{1i} + U_{nb}(r_{1i})) \qquad (3.22)$$

where U_{nb} is defined in eq. (3.11). With this model potential, run MD simulations, and calculate the free energy associated with changing the ion charge from 0 to +1.

Solution 3.5. Use Program 3.B.

3.4. COMBINING SOLVENT EFFECTS WITH QUANTUM MECHANICAL SOLUTE CALCULATIONS

Now knowing how to evaluate solvation-free energies, we are ready to explore the effect of the solvent on the potential surface of the reacting solute atoms. Adapting the EVB approach we can describe the reaction by including the solute–solvent interaction in the diagonal elements of the solute Hamiltonian, using

$$H_{ii} = H_{ii}^0 + U_{Ss}^i + U_{ss} + U_{ind}^{(i)}$$

$$H_{ij} = H_{ij}^0 \tag{3.23}$$

Here (in contrast to the approach taken in Chapter 2) we do not assume that the energy of each valence bond structure is correlated with its solvation-free energy. Instead we use the actual ground-state potential surface to *calculate* the ground-state free energy. To see how this is actually done let's consider as a test case an S_N2 type reaction which can be written as

$$X^- + CH_3Y \rightarrow XCH_3 + Y^- \tag{3.24}$$

This reaction can be described by the three resonance structures

$$\Phi_1 = [X^-C\text{–}Y\ ]\chi_1$$

$$\Phi_2 = [X\text{–}C\ Y^-]\chi_2$$

$$\Phi_3 = [X^-C^+Y^-]\chi_3 \tag{3.25}$$

The relevant Hamiltonian for the gas-phase solute molecules can be treated by the same three-orbitals four-electron model used in Chapter 2. Since the energy of Φ_3 is much higher than that of Φ_1 and Φ_2 (see Table 2.4), we represent the system by its two lowest energy resonance structures, using now the notation ψ_1 and ψ_2 as is done in eq. (2.40). The energies of these two effective configurations are now written as

$$\varepsilon_1^0 = H_{11}^0 = \Delta M(b_1) + U_{nb}^{(1)} + U_{ind}^{(1)} + \frac{1}{2} \sum_m K_{b,m}^{(1)} (b_m^{(1)} - b_{0,m}^{(1)})^2$$

$$+ \frac{1}{2} \sum_m K_{\theta,m}^{(1)} (\theta_m^{(1)} - \theta_{0,m}^{(1)})^2$$

$$\varepsilon_2^0 = H_{22}^0 = \Delta M(b_2) + U_{nb}^{(2)} + U_{ind}^{(2)} + \alpha_2^0 + \frac{1}{2} \sum_m K_{b,m}^{(2)} (b_m^{(2)} - b_{0,m}^{(2)})^2$$

$$+ \frac{1}{2} \sum_m K_{\theta,m}^{(2)} (\theta_\theta^{(2)} - \theta_{0,m}^{(2)})^2$$

$$H_{12} = A \exp\{-\mu(r_3 - r_3^0)\} \tag{3.26}$$

in which the gas-phase energies ε_i^0 are approximated by analytical potential functions, similar to those used in Chapter 2. b_1, b_2, and r_3 are the X–C, C–Y, and X–Y distances, the b_m's are the C–H bond lengths, and the θ_m's are the X–C–H, H–C–H, and H–C–Y angles defined by the covalent bonding arrangement for a given resonance structure. For example, in ψ_1 the θ_m is defined by the H–C–Y and H–C–H angles, while the K_θ for the X–C–H angles is set to zero. The U_{nb}^i are the nonbonded interactions between the nonbonded atoms in the ith resonance structure. For ψ_1 this includes the $X^- \cdots C$ repulsion, the $X^- \cdots H$ repulsion and the $X^- \cdots Y$ interaction. These nonbonded interactions are described by either $Ae^{-\mu r}$ or 6–12 van der Waals potential functions (see Table 3.3). The induced-dipole terms describe the attraction between the charges of the system (Q) and its induced dipoles. This term is evaluated for simplicity by neglecting the interactions between induced dipoles. Finally, the α_2^0 parameter is the energy difference between ψ_2 and ψ_1 with the fragments at infinite separation (this is simply the difference in the free energy of formation of $X^- + CH_3Y$ and $Y^- + CH_3X$). As before, we can determine the parameters in eq. (3.26) by fitting the corresponding ground-state potential surface to ab initio calculations and relevant experimental information. A reasonable parameter set is given in Table (3.3).

TABLE 3.3. Parameters for the Potential Functions of the $Cl^- + CH_3Cl \rightarrow ClCH_3 + Cl^-$ S_N2 Reaction

Bonds		$U_b = \frac{1}{2} K_b (b - b_0)^2 + D[1 + \exp\{-a(b - b_0)\}]^2$		
C–Cl	$K_b = 0$	$D = 60$	$b_0 = 1.8$	$a = 2.0$
C–H	$K_b = 310$	$D = 0$	$b_0 = 1.102$	$a = 0.0$

Bond Angles		$U_\theta = \frac{1}{2} K_\theta (\theta - \theta_0)^2$	
H–C–H		$K_\theta = 72$	$\theta_0 = 1.911$
H–C–Cl		$K_\theta = 60$	$\theta_0 = 1.911$

Nonbonded		$U_{nb} = A_1 \exp\{-ar\} + A_2 r^{-12} + Br^{-6}$		
$C \cdots Cl^-$	$A_1 = 0$	$A_2 = 1.39 \times 10^5$	$B = 285$	$a = 0$
$Cl \cdots Cl^-$	$A_1 = 0$	$A_2 = 1.26 \times 10^{12}$	$B = 8305$	$a = 0$
$Cl^- \cdots H$	$A_1 = 11297$	$A_2 = 0$	$B = 120$	$a = 3.6$

Nonbonded	$U'_{nb} = A_i A_j r^{-12} - B_i B_j r^{-6}$	
H	$A = 5.0$	$B = 2.0$
O	$A = 793$	$B = 26$
Cl^-	$A = 2000$	$B = 2.5$
Cl	$A = 1500$	$B = 2.5$
C	$A = 632$	$B = 20$

Charges		$U_{qq} = 332 q_i q_j / r_{ij}$		
$(Cl^-_{(1)}CH_3 - Cl_{(2)})(\psi_1)$	$q_{Cl^-_{(1)}} = -1$	$q_C = 0$	$q_H = 0$	$q_{Cl_{(2)}} = 0$
$(Cl_{(1)} - CH_3Cl^-_{(2)})(\psi_2)$	$q_{Cl_{(1)}} = 0$	$q_C = 0$	$q_H = 0$	$q_{Cl^-_{(2)}} = -1$

Off-Diagonal Parameters			
H_{12}	$A_{12}^{ClCl} = 3256$	$r^0 = 0.0$	$\mu = 1.0$

[a]Energies in kcal/mol, distances in Å, angles in radians, and charges in atomic charge units.
[b]U_{nb} describes the nonbonded interactions between the solute atoms, while U'_{nb} is used for the nonbonded interactions between the solute and the solvent. The solvent charges are taken as −0.82 and 0.41 for O and H respectively.

With the gas-phase potential surface we can obtain the solution Hamiltonians by eq. (3.23), adding the solvent–solute interaction to the "classical" part of the diagonal EVB matrix elements. That is, we use

$$\varepsilon_1 = H_{11} = H_{11}^0 + U_{Qq,Ss}^{(1)} + U_{nb,Ss}^{(1)} + U_{ind}^{(1)} + U_{ss}$$

$$\varepsilon_2 = H_{22} = H_{22}^0 + U_{Qq,Ss}^{(2)} + U_{nb,Ss}^{(2)} + U_{ind}^{(2)} + U_{ss}$$

$$H_{12} = H_{12}^0 \tag{3.27}$$

where S and s designate solute and solvent, respectively. U_{Qq} is the electrostatic interaction between the solute charges ($\mathbf{Q}$) in the given resonance structure and the solvent residual charges ($\mathbf{q}$). U_{nb} is the nonbonded interaction between the solute and solvent atoms, while U_{ind} is the inter-

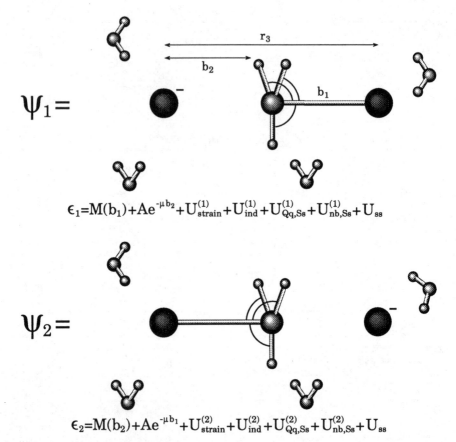

$$\epsilon_1 = M(b_1) + Ae^{-\mu b_2} + U_{strain}^{(1)} + U_{ind}^{(1)} + U_{Qq,Ss}^{(1)} + U_{nb,Ss}^{(1)} + U_{ss}$$

$$\epsilon_2 = M(b_2) + Ae^{-\mu b_1} + U_{strain}^{(2)} + U_{ind}^{(2)} + U_{Qq,Ss}^{(2)} + U_{nb,Ss}^{(2)} + U_{ss}$$

FIGURE 3.3. A schematic description of the two low-energy resonance structures used to describe the S_N2 reaction [see eq. (3.26) for more details].

action between the induced dipoles of the solvent and the solute charges. The two resonance structures and the corresponding force fields are described schematically in Fig. 3.3.

Exercise 3.6. Write a program that evaluates ε_1, ε_2, and H_{12} for the S_N2 reaction $Cl^- Cl–Cl \rightarrow Cl–C \ Cl^-$, neglecting the hydrogens on the carbon so that the θ term in eq. (3.26) is not needed. Also, neglect the U_{ind} term. Next, surround this "solute" system by 20 dipoles and simulate the resulting solute + solvent system with the potential $U = \varepsilon_1$ (examine the distances between the three atoms during the simulation).

Solution 3.6. See Program 3.C.

3.5. EVALUATION OF ACTIVATION-FREE ENERGIES

3.5.1. The EVB Mapping Potential

Once the analytical ground-state potential surface is constructed, we can start to explore the configuration space and evaluate the reaction free-energy surface and the activation free energy, $\Delta g^{\neq}$. The strategy for this involves the free-energy perturbation approach described above, but now we use a mapping potential which is composed of the EVB diagonal energies and is given by

$$\varepsilon_m = \varepsilon_1(1 - \lambda_m) + \varepsilon_2 \lambda_m \qquad (3.28)$$

This mapping potential [which is equivalent in many respects to the potential used in eq. (3.17)] can drive the system through the "non physical" process of transforming ψ_1 to ψ_2, which can be considered as changing the "molecule" $X^- CH_3Y$ to $XCH_3 \ Y^-$. Note that ε_1 has a minimum at the reactant geometry and ε_2 has a minimum at the product geometry. Thus, as λ_m is changed from zero to one, the system is forced to change from the reactant state to the product state. In particular, the solvent is forced to adjust its polarization to the changing charge distribution on the solute.

The free energy associated with changing ε_1 to ε_2 can be obtained by eq. (3.20), by a complete analogy to the adiabatic charging procedure described in Section 3.3. The free-energy function $\Delta G(\lambda)$ reflects both the electrostatic (solvation) effects associated with the changes of the solute charges, as well as the intramolecular effects associated with changing ε_1 to ε_2. Figure 3.4 demonstrates the dependence of $\Delta G(\lambda)$ on λ for an exchange reaction where $X = Y$. The figure gives both the total $\Delta G(\lambda)$ and its electrostatic components.

Obtaining $\Delta G(\lambda)$ with the perturbation procedure described above and using the mapping potentials ε_m is not sufficient for evaluating the activation

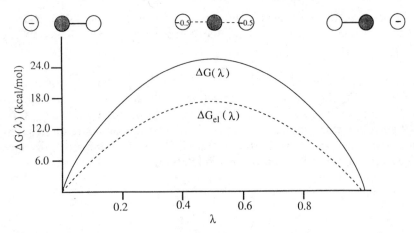

FIGURE 3.4. The dependence of the mapping free energy for an S_N2 reaction on the parameter λ. The figure shows both the total free energy $\Delta G(\lambda)$ (solid line) and the electrostatic contribution $\Delta G_{el}(\lambda)$ (dotted line), demonstrating that the largest contribution to $\Delta G(\lambda)$ is due to electrostatic interactions (Ref. 11).

free energy, $\Delta g^{\neq}$, which reflects the probability of being at the transition state on the actual ground-state potential surface. The function $\Delta G(\lambda)$ is basically used to find the free energy $\Delta G(\lambda^{\neq})$ associated with changing ε_1 to a potential $\varepsilon_{m^{\neq}}$ (e.g., $\lambda^{\neq} = 0.5$ in the example of Fig. 3.4), that forces the system to spend the longest time near the actual transition state. However, one still needs to find the probability of reaching the transition state for trajectories that move on the actual ground state, E_g. The corresponding procedure is described below.

3.5.2. Obtaining the Free-Energy Functions

In order to obtain the probability of being at the transition state, we have to define first a reaction coordinate for our system. A general reaction coordinate X (see eq. 2.9) can be defined in terms of the energy gap $\varepsilon_2 - \varepsilon_1$ (Ref. 11). This is done by dividing the configuration space of the system into subspaces s'' that satisfy the relationship $\Delta\varepsilon(s'') = X''$, in which the energy differences X'' are constants and $\Delta\varepsilon = \varepsilon_2 - \varepsilon_1$. The parameter X'' is then used as the reaction coordinate and the corresponding free energy, $\Delta g(X'')$, is given by Ref. 11

$$\exp\{-\Delta g(X'')\beta\} \cong \exp\{-\Delta G(\lambda_m)\beta\}\langle\exp\{-(E_g(X'') - \varepsilon_m(X''))\beta\}\rangle_m$$

$$(3.29)$$

This expression relates the probability of find the system at X'' on the ground state E_g to the probability of being on the mapping potential ε_m (that keeps X around X''). The evaluation of $\Delta g(X'')$ for an exchange reaction is described in Fig. 3.5.

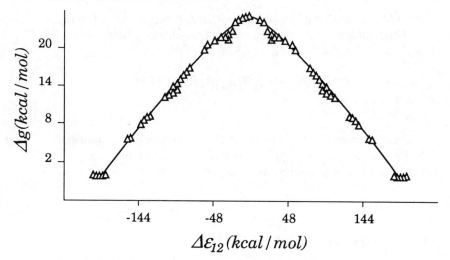

FIGURE 3.5. The actual free-energy profile for the ground-state surface as a function of the energy gap $\Delta\varepsilon$. The calculations are done for the $Cl^- + CH_3Cl \rightarrow ClCH_3 + Cl^-$ exchange reaction (Ref. 11).

Exercise 3.7. Consider the system introduced in Exercise 3.6. Calculate both the free energy of moving from ε_1 to ε_2 $[\Delta G(\lambda)]$ and the reaction free energy $\Delta g(\Delta\varepsilon)$.

Solution 3.7. Use program 3.C.

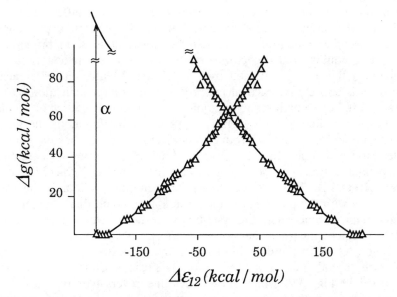

FIGURE 3.6. The free-energy functionals Δg_1 and Δg_2 as a function of the energy gap $\Delta\varepsilon$ for the $Cl^- + CH_3Cl \rightarrow ClCH_3 + Cl^-$ reaction (Ref. 11).

In addition to $\Delta g(X^n)$, we can obtain the probability of being at X^n on ε_i. This probability defines the free-energy functions ($\Delta g_i(X^n) = \int \exp\{-\varepsilon_i(X^n)\beta\} \, ds^n$) given by (Ref. 11)

$$\exp\{-\Delta g_i(X^n)\beta\} = \exp\{-\Delta G(\lambda_m)\beta\}\langle\exp\{-(\varepsilon_i(X^n) - \varepsilon_m(X^n)\beta\}\rangle_m$$

(3.30)

The evaluation of the Δg_i's for an exchange reaction is demonstrated in Fig. 3.6. The Δg_1 and Δg_2 curves can be used to estimate the dependence of $\Delta g^{\neq}$ on the reaction free energy ΔG_0, as will be shown in Section 3.7.

3.6. EXAMINING DYNAMICAL EFFECTS

Following Chapter 2 we write the rate constant of eq. (2.11) as

$$k = \bar{\tau}^{-1} \exp\{-\Delta G^{\neq}\beta\} \simeq \bar{\tau}^{-1} \exp\{-\Delta g^{\neq}\beta\}$$

(3.31)

where $\Delta g^{\neq}$ denotes the $\Delta g(X^{\neq})$ of eq. (2.11) and the factor ($\Delta X^{\neq}/\int_{-\infty}^{X^{\neq}} \exp\{-\Delta g\beta\} \, dX$) of eq. (2.11) is approximated here by unity (this is a reasonable approximation in many cases, while problems where this factor must be evaluated will be addressed in Chapter 9). As in Chapter 2 we take $\bar{\tau}$ as the average time needed for a productive trajectory to pass the transition state region. This time is around 10^{13} sec^{-1} at room temperature, but it may have somewhat different values in solvents of different viscosity. In order to examine possible dynamical effects it is important to evaluate $\bar{\tau}$ by microscopic simulations. This can be done, at least in principle, by generating equilibrated trajectories of the system at the reactant state and evaluating the velocity of reactive trajectories. Unfortunately, such an approach is entirely impractical for processes with large activation barriers, because it takes an extremely long time until the kinetic energy of the many degrees of freedom of the system is converted to motion along the reaction coordinate and generates a productive trajectory that reaches the transition state. However, one can use a practical trick by preparing an equilibrated system at the transition state region and then running "*downhill*" trajectories (Ref. 12). This can be done, for example, by letting the system equilibrate with the potential $\varepsilon^{\neq}$ mentioned above, and then changing $\varepsilon^{\neq}$ to E_g at a point in time where the system crosses the transition state (i.e., when $\varepsilon_1 = \varepsilon_2$). Since classical dynamics is invariant to time reversal, we can use the time reversal of the downhill trajectories to construct the corresponding very rare "uphill" productive trajectory. The results of such a procedure are described in Fig. 3.7, which shows the time reversal of typical downhill trajectories for an S_N2 reaction.

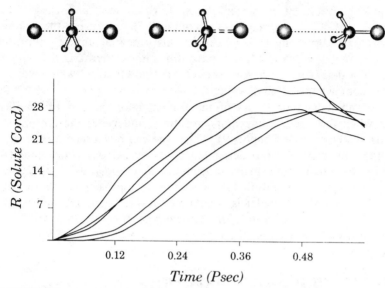

FIGURE 3.7. Showing the time-dependent solute coordinate R in a typical downhill trajectory for the S_N2 reaction $Cl^- + CH_3Cl \rightarrow ClCH_3 + Cl^-$. Such a class of trajectories gives the average forward velocity $(\Delta R_+/\Delta t)$ at the transition state $(R = 0)$. The average time required for the trajectories to move from $R = R^{\neq} = 0$ to $R^{\neq} + \Delta R^{\neq}$ (where the potential energy at $R^{\neq} + \Delta R^{\neq}$ is lower than that at $R^{\neq}$ by β^{-1}) is our $\bar{\tau}$. Note that a more rigorous result will be obtained by monitoring $\varepsilon_2 - \varepsilon_1$ rather than R (see Ref. 11).

The downhill trajectory approach is useful not only for evaluating the preexponential factor $\bar{\tau}$, but also for examining the relationship between the dynamics of the solute and solvent contributions to the reaction coordinate. That is, as discussed in detail in Ref. 11, the reaction coordinate can be described in terms of the solute and the solvent reaction coordinates, where the solvent coordinate [which corresponds to the solvent contribution to the energy gap $(\varepsilon_2 - \varepsilon_1)$] is related to the solvent polarization. A reaction can be considered as *solvent driven reactions* if the solute coordinate changes towards its transition state value, only after the solvent reaches configurations whose polarization stabilizes the transition state charge distribution. On the other hand a reaction can be considered as a *solute driven reaction* if the solvent polarization changes to its transition state value, only after the solute coordinate and charge distribution change to their transition state values (see Fig. 11 of Ref. 11a). Downhill trajectories allow one to explore the time dependence of the reactive trajectories and to examine in detail the individual dynamics of the solute and solvent coordinates (see Ref. 11). Other related approaches are discussed in Ref. 18.

Exercise 3.7. Use the system of Exercise 3.6 (the three atoms Cl^- C–Cl system) and generate a downhill trajectory, starting with $b_1 = 1.6$ Å, $b_2 =$

1.8 Å, and the potential surface E_g. Estimate the time $\bar{\tau}$, as discussed in the caption of Fig. 3.7. You may also accomplish this by using the potential $\varepsilon^{\neq} = (\varepsilon_2 + \varepsilon_1)/2$, running trajectories on this potential until equilibration, and then changing $\varepsilon^{\neq}$ to E_g. If you have difficulties generating the potential E_g or its first derivatives, use for simplicity the potential ε_1 (which will give, however, somewhat too small value for $\bar{\tau}$).

The possible dynamical role of the solvent coordinate is of significant interest when one studies enzymatic reactions and views the enzyme as a generalized solvent (see Chapter 9). This issue can be studied effectively by estimating the average time dependence of the solvent coordinate (that would be obtained from many downhill trajectories) using the *Linear response approximation* (Ref. 11). This useful approach is not discussed here and the interested reader is encouraged to read Ref. 11b and Ref. 19. The corresponding expression for the average time dependence of the solvent coordinate will be given and used in Chapter 9.

3.7. LINEAR FREE-ENERGY RELATIONSHIPS

Consideration of the relationship between the ground state free energy Δg of Fig. 3.5, and the functions Δg_1 and Δg_2 of Fig. 3.6 gives an interesting insight about the correlation between the reaction free energy ΔG_0 and the activation free energy $\Delta g^{\neq}$. That is, using the solution of the 2×2 Hamiltonian of eq. (3.27) $(E_g = \frac{1}{2}[(\varepsilon_1 + \varepsilon_2) - ((\varepsilon_2 - \varepsilon_1)^2 + 4H_{12}^2)^{\frac{1}{2}}])$ at the transition state where $\varepsilon_1 = \varepsilon_1$ and at the ground state minimum; X_0, where $(\varepsilon_2 - \varepsilon_1) \gg H_{12}^2$, gives the relationships $E_g(X^{\neq}, s) \simeq \varepsilon_1(X^{\neq}, s) - H_{12}(X^{\neq}, s)$ and $E_g(X_0, s) \simeq H_{12}^2(X_0, s)/[\varepsilon_2(X_0, s) - \varepsilon_1(X_0, s)]$ where s is the coordinate perpendicular to our X [see eq. (2.9)]. Since $\Delta g(X)$ is related to the thermal average of $E_g(X, s)$ over the s space, we can use the approximation

$$\Delta g^{\neq} = \Delta g(X^{\neq}) - \Delta g(X_0)$$
$$\cong \Delta g_1(X^{\neq}) - \Delta g_1(X_0) - \bar{H}_{12}(X^{\neq}) + \langle H_{12}^2(X_0)/[(\varepsilon_2(X_0) - \varepsilon_1(X_0)] \rangle_s$$

(3.32)

where $\bar{H}_{12}(X^{\neq})$ is the average of H_{12} over s at $X^{\neq}$ and $\langle \ \rangle_s$ designates an average over s. We can further manipulate eq. (3.32) by assuming that Δg_1 and Δg_2 can be approximated by paraboli with the same curvatures (see Fig. 3.8). With this assumption, we can evaluate the Δg at the intersection of the two paraboli by

$$\Delta g_1(X^{\neq}) = (\Delta G_0 + \alpha)^2/4\alpha$$

(3.33)

where α, which is called the *reorganization energy*, (see Fig. 3.8), is given by the difference in the value of Δg_2 at the minimum of Δg_1 (the reactant state)

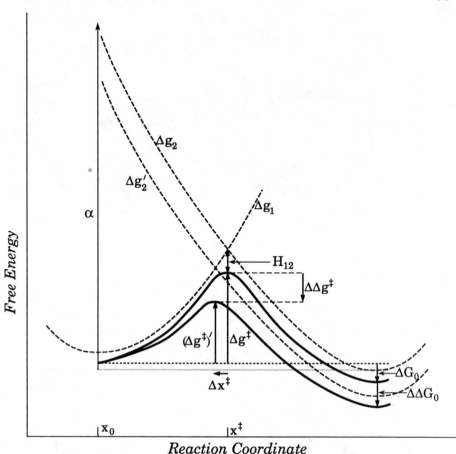

Reaction Coordinate

FIGURE 3.8. A schematic description of the relationship between the free-energy difference ΔG_0 and the activation free energy $\Delta g^{\neq}$. The figure illustrates how a shift of Δg_2 by $\Delta\Delta G_0$ (that changes Δg_2 to $\Delta g_2'$ and ΔG_0 to $\Delta G_0 + \Delta\Delta G_0$) changes $\Delta g^{\neq}$ by a similar amount.

and the minimum of Δg_2 (the product state). The reader can easily verify eq. (3.33) in the simple case where $\Delta G_0 = 0$ and $\Delta g_1(X^{\neq}) = \alpha/4$. With eq. (3.32) we can now write the actual activation free energy as

$$\Delta g^{\neq} \approx (\Delta G_0 + \alpha)^2/4\alpha - \bar{H}_{12}(X^{\neq}) + \bar{H}_{12}^2(X_0)/\alpha$$

$$|\Delta G_0| < \alpha \tag{3.34}$$

Outside the range of ($|\Delta G_0| < \alpha$) we have the situation shown in Fig. 3.9, so that $\Delta g^{\neq}$ can be approximated by

$$\Delta g^{\neq} \approx \Delta G_0$$

$$\Delta G_0 > \alpha \tag{3.35}$$

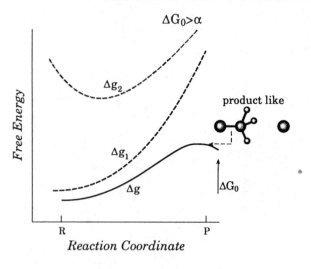

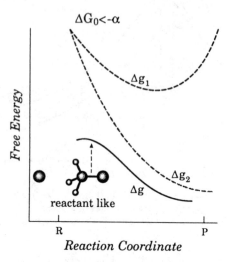

FIGURE 3.9. A schematic description of the free-energy functionals at the range where $|\Delta G_0| > \alpha$.

and
$$\Delta g^{\neq} \approx 0$$

$$\Delta G_0 < -\alpha \qquad (3.36)$$

Equation (3.34) without the $H_{12}(X^{\neq})$ and the $\bar{H}_{12}^2/\alpha$ terms is identical to the Marcus' equation for methyl transfer reactions (Ref. 13). This equation predicts, at the range $(|\Delta G_0| < \alpha)$, a linear relationship between $\Delta\Delta G_0$ and $\Delta\Delta g^{\neq}$ by

$$\Delta\Delta g^{\neq} \approx \theta \, \Delta\Delta G_0 = [(\Delta G_0 + \alpha)/2\alpha] \, \Delta\Delta G_0 \qquad (3.37)$$

where θ is the correlation coefficient between ΔG_0 and $\Delta g^{\neq}$. Linear free-energy relationships, which date back to Brönsted (Ref. 14) and Hammond (Ref. 15), have played a major role in physical organic chemistry (Refs. 16 and 17). The problem is, however, that such relationships, including eq. (3.34), are not exact nor based on a fundamental microscopic principle. In fact, the Δg curves will not be harmonic if the solvent response to the solute charges does not obey the linear response approximation [the actual behavior of these Δg's can only be explored by microscopic considerations (Ref. 11)]. Furthermore, the parameter α is not expected to be reproduced accurately by macroscopic estimates. Nevertheless, if α can be estimated from microscopic calculations, then even an approximate validity of eq. (3.34) can provide a powerful way of avoiding the actual evaluation of $\Delta g^{\neq}$.

A microscopic examination of the linear free-energy relationship for $S_N 2$ reactions is described in Fig. 3.10. The figure presents the calculated

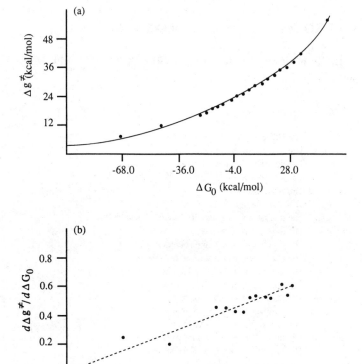

FIGURE 3.10. (a) Showing the relationship between the activation free energy $\Delta g^{\neq}$ and the reaction free energy ΔG_0 for the $X^- + CH_3 Y \rightarrow XCH_3 + Y^-$ system. (b) The dependence of the "linear" correlation coefficient $\theta = d \Delta g^{\neq} / d \Delta G_0$ on ΔG_0.

dependence of $\Delta g^{\neq}$ on ΔG_0 for a hypothetical system where all the potential parameters are set at their values for the $Cl^- + CH_3Cl \rightarrow ClCH_3 + Cl^-$ reaction except the α_2^0 of eq. (3.26), which is varied in a parametric way. This corresponds to a variation of the electronegativity of a Cl^- type atom while keeping its radius constant. The calculated results support the approximate validity of eqs. (3.34), (3.35) and (3.36), where θ changes from around one for very endothermic reactions with a *product-like* transition state, to 0.5 for $\Delta G_0 \sim 0$ [see eq. (3.37)], and then to zero [i.e., eq. (3.36)] for very exothermic reactions with *reactant-like* transition states. The qualitative trend of eq. (3.34) is also supported by many experimental studies (Ref. 16). However, the experimental evidences are not conclusive since it is hard to find experimental systems where only α and ΔG_0 change without a change of H_{12} (which also affects $\Delta g^{\neq}$). Such an experiment, however, can easily be performed on a computer, as is demonstrated in this chapter (see also Ref. 11).

3.8.   SOME RELEVANT COMPUTER PROGRAMS

3.A.   Evaluations of the Radial Distribution Function by the MD Simulations

```
c         Molecular dynamics of a dipolar solvent with two atoms per solvent molecule.

c         crg         The charges of the solvent atoms.
c         crgs        The charges of the solute atoms.
c         dt          The time step in units of 0.049 pico seconds.
c         ekin        The kinetic energy.
c         epot        The potential energy.
c         etot        The total energy.
c         fl          The mapping parameters for free energy calculations.
c         fmass       The masses of the atoms .
c         fmu         The dipole moment of the solvent molecules.
c         iac         The type of atoms for the solute molecule.
c         nd          The number of  solvent molecules.
c         nr          The number of solute resonance structures.
c         ns          The number of solutes atoms.
c         nsolv       The number of solvent atoms.
c         rad         The radius of the solvent sphere.
c         temp        The target temperature of the system.
c         temp1       The current temperature of the system.
c         xs          The solute coordinates.
          implicit real*8(a-h,o-z)
          parameter (ndip=500,natom=2*ndip,ncoord=3*natom)
          common/param/crgs(20,4),crg(2),avdw(8),bvdw(8),fmass(8),
     &                     rz(8),am(8),d(8),r12,dt,temp,rad
          common/cntrl/iter,niter,epot,ekin,etot,temp1
          common/coords/nd,ns,nsolv,nr,iac(20,4),x(3,natom),xs(3,20),
     &                     nb(10),ib(10,4),jb(10,4)
```

```
          common/mapping/fl(4),emap(4,10000),imap
          call readin
          fl(1)=1.0                                    ! Equilibration run.
          fl(2)=0.0
          iter=0
          call trajec
          call map
          end
c=========================================================================
          subroutine map

c         This subroutine performs the free energy perturbation calculations.
c         Using the mapping parameter lamda that drives the system from
c         its initial to final state in nmap steps.

c         emap(1,ll)  The potential associated with the neutral atoms.
c         emap(2,ll)  The potential associated with the charged atoms.
c         e1          The mapping potential for lamda(l).
c         e2          The mapping potential for lamda(l+1).
          implicit real*8(a-h,o-z)
          parameter (ndip=500,natom=2*ndip,ncoord=3*natom)
          common/param/crgs(20,4),crg(2),avdw(8),bvdw(8),fmass(8),
         &                rz(8),am(8),d(8),r12,dt,temp,rad
          common/cntrl/iter,niter,epot,ekin,etot,temp1
          common/coords/nd,ns,nsolv,nr,iac(20,4),x(3,natom),xs(3,20),
         &                nb(10),ib(10,4),jb(10,4)
          common/mapping/fl(4),emap(4,10000),imap
          real * 8 lamda
          nmap=10                          ! Here we perform nmap MD simulations, gradually
          imap=0                           ! changing lamda and the corresponding mapping
          delta=1./(nmap+1)                ! potential e(l), e(l)=((1-lamda(l))*U1+lamda(l)*U2),
          do lmap=2,nmap                   ! where U1 and U2 are stored in the emap(1) and
             lamda=(lmap-1)*delta          ! emap(2) arrays.
             fl(1)=1.0-lamda
             fl(2)=lamda
             call trajec
             write(6,*) 'l m a p',lmap
          enddo
          beta=1./(0.6*temp/300)           ! The mapping free energy is evaluated below.
          dg=0.
          ll=1                             ! With the precalculated values of U1 and U2
          do 20 l=2,nmap                   ! (emap(1) and emap(2)) we evaluate the free
             sume=0.                       ! energy difference deltaG ((lamda(1)--> lamda(l+1)).
             sums=0.
             do i=1,niter
                ll=ll+1
                lamda=(l-1)*delta
                fl(1)=1.0-lamda
                fl(2)=lamda
                e1=emap(1,ll)*fl(1)+emap(2,ll)*fl(2)
                e2=emap(1,ll)*(fl(1)-delta)+emap(2,ll)*(fl(2)+delta)
                dv=e2-e1
                if(mod(i,20).eq.1) write(6,*) 'dv',dv
                sume=sume+dexp(-dv*beta)
                sums=sums+1
```

```
                enddo
                dg1=-log(sume/sums)/beta
                dg=dg+dg1
                write(6,*) 'delta g ',dg
20          continue
            end
c=============================================================
            subroutine readin

c       fm          The mass of the atoms.
c       ib,jb       The list of bonded atoms.
c       nb          The number of bonds.
c       x           The coordinates of the solvent atoms.
            implicit real*8(a-h,o-z)
            parameter (ndip=500,natom=2*ndip,ncoord=3*natom)
            common/cntrl/iter,niter,epot,ekin,etot,temp1
            common/coords/nd,ns,nsolv,nr,iac(20,4),x(3,natom),xs(3,20),
        &                  nb(10),ib(10,4),jb(10,4)
            common/dynmcs/fm(natom),vn(ncoord),anm1(ncoord),an(ncoord)
            common/param/crgs(20,4),crg(2),avdw(8),bvdw(8),fmass(8),
        &                  rz(8),am(8),d(8),r12,dt,temp,rad
            data fmass/1.,16.,14.,12.,35.,43.,23.,39./
            read(5,*) tspan,dt,temp,rad,fmu
            write(6,*) 'rad',rad
            read(5,*) crg(1)
            read(5,*) crg(2)
            r12=dabs(fmu/crg(1))
            read(5,*) ns
            write(6,*) 'ns',ns
            if(ns.ne.0) then
                read(5,*) nr
                write(6,*) 'nr',nr
                do ii=1,nr
                    read (5,*) nb(ii)
                    nb1=nb(ii)
                    do i=1,nb1
                        if(nb1.ne.0) read (5,*) ib(i,ii),jb(i,ii)
                    enddo
                    read(5,*) (iac(i,ii),i=1,ns)
                    read(5,*) (crgs(i,ii),i=1,ns)
                    write(6,*) 'crgs', (crgs(i,ii),i=1,ns)
                enddo
                do i=1,ns
                    read(5,*) (xs(k,i),k=1,3)
                enddo
            endif
            icont=0
            if (icont.eq.0) call grid
            nsolv=2*nd
            write(6,*) 'nsolv',nsolv
            do i=1,ns
                do k=1,3
                    if(ns.ne.0) x(k,i+nsolv)=xs(k,i)
                enddo
```

```
        enddo
        ntot=2*nd+ns
        do i=1,nd
              fm(i*2-1)=fmass(1)
              fm(i*2)=fmass(2)
        enddo
        do i=nd+1,ntot
              if(ns.ne.0) fm(i)=fmass(iac(i,1))
        enddo
        print '(a,i4,a)',' The system contains',nd,' dipoles.'
        niter=400
        print '(a,i6)',' Number of iterations to be performed: ',niter
        return
        end
c=================================================================
        subroutine g_r

c       This subroutine calculates the three radial distribution functions for the
c       solvent. The radial distribution functions provide information on the
c       solvent structure. Specially, the function g-AB(r) is the average number
c       of type B atoms within a spherical shell at a radius r centered on an
c       arbitary type A atom, divided by the number of type B atoms that one
c       would expect to find in the shell based on the bulk solvent density.

c       del       1.0/dr
c       ghist     The radial distribution function.
c       lmax      The maximum value of l.
c       nhist     The actual number of atoms found within the
c       spherical shell extending from r to r+dr about a central atom.
        implicit real*8(a-h,o-z)
        parameter (ndip=500,natom=2*ndip,ncoord=3*natom)
        real*8 dr(3)
        common/cntrl/iter,niter,epot,ekin,etot,temp1
        common/coords/nd,ns,nsolv,nr,iac(20,4),x(3,natom),xs(3,20),
     &                nb(10),ib(10,4),jb(10,4)
        common/mapping/fl(4),emap(4,10000),imap
        common/param/crgs(20,4),crg(2),avdw(8),bvdw(8),fmass(8),
     &                rz(8),am(8),d(8),r12,dt,temp,rad
        common/rdf/ghist(3,250),nhist(3,250),nrdf
        data nhist/750*0/,nrdf/0/
        del=1.d0
        lmax=1+1.5d0*del*rad
        do 10  i=1,nsolv-2                      ! Here we run over the solvent molecules which
              mi=1+mod(i-1,2)                   ! are composed of two atoms (A and B) each.
              id=(i+1)/2
              do 20  j=i+1,nsolv
                   mj=1+mod(j-1,2)              ! We only wish to collect intermolecular terms in the
                   jd=(j+1)/2                   ! histograms, so we multiply by the filter inc, where
                   inc=jd-id                    ! inc=1 for atoms in different molecules, and inc=0
                   inc=min0(inc,1)              ! for atoms in the same molecule.
                   k=mi+mj-1                    ! All three radial distribution functions are calculated
                   dr(1)=x(1,i)-x(1,j)          ! within this loop; k={1,2,3} correspond to the
                   dr(2)=x(2,i)-x(2,j)          ! functions {g-AA, g-AB,g-BB} respectively.
```

```
                    dr(3)=x(3,i)-x(3,j)
                    r2=dr(1)*dr(1)+dr(2)*dr(2)+dr(3)*dr(3)
                    l=del*dsqrt(r2)
                    l=min0(l,lmax)
                    nhist(k,l)=nhist(k,l)+inc*2
    20          continue
    10      continue
            nrdf=nrdf+1
            if (mod(iter,80).ne.1) return
            s=(del*rad)**3/(nd*nd*nrdf)
            do 30 l=1,lmax-1
                fltl=l
                rl=fltl/del
                y=rl/rad
                f=16.d0/((y-2.d0)*(y-2.d0)*(y+4.d0))          ! A scaling correction for spherical systems.
                ghist(1,l)=f*s*nhist(1,l)/(3.d0*fltl*(fltl+1.d0)+1)
                ghist(2,l)=0.5d0*f*s*nhist(2,l)/(3.d0*fltl*(fltl+1.d0)+1)
                ghist(3,l)=f*s*nhist(3,l)/(3.d0*fltl*(fltl+1.d0)+1)
    30      continue
            write (6,*) 'radial distribution function'
            print '(10i7)',(l,l=1,lmax-1)
            print '(10f7.3)',(ghist(1,l),l=1,lmax-1)
            return
            end
c=====================================================================
        subroutine trajec

c       This subroutine controls the number of MD steps.

        implicit real*8(a-h,o-z)
        common/cntrl/iter,niter,epot,ekin,etot,temp1
        common/param/crgs(20,4),crg(2),avdw(8),bvdw(8),fmass(8),
     &                rz(8),am(8),d(8),r12,dt,temp,rad
        do iter=1,niter
            call beeman
            etot=epot+ekin
            if(mod(iter,20).eq.1) then
                print 100 ,epot,ekin,etot
                print 101 ,temp,temp1
            endif
            call g_r                                           ! Calculates the radial distribution functions.
        enddo
        return
    100 format(2x,'epot,ekin,etot',3f10.2)
    101 format(2x,'temp ',2x,f10.2,'temp1 ',f10.2)
        end
c=====================================================================
        subroutine beeman

c       3rd-order beeman numerical integration algorithm.

c       x1       The coordinates.
c       vn       The velocity times dt.
c       an       The current acceleration times dtt.
```

```
c        anm1     The previous acceleration times dtt.
         implicit real*8(a-h,o-z)
         parameter (ndip=500,natom=2*ndip,ncoord=3*natom)
         common/cntrl/iter,niter,epot,ekin,etot,temp1
         common/coords/nd,ns,nsolv,nr,iac(20,4),x(3,natom),xs(3,20),
     &                 nb(10),ib(10,4),jb(10,4)
         real*8 x1(ncoord),d1(ncoord)
         common/dynmcs/fm(natom),vn(ncoord),anm1(ncoord),
     &                 an(ncoord)
         common/force/du(3,natom),dx(3)
         common/param/crgs(20,4),crg(2),avdw(8),bvdw(8),fmass(8),
     &                 rz(8),am(8),d(8),r12,dt,temp,rad
         equivalence (x(1,1),x1(1)),(du(1,1),d1(1))
         dtt=dt*dt/6.d0
         nx=6*nd+3*ns
         do i=1,nx
             x1(i)=x1(i)+vn(i)+4.d0*an(i)-anm1(i)
             vn(i)=vn(i)+5.d0*an(i)-anm1(i)
             anm1(i)=an(i)
         enddo
         call energy
         ekin=0.d0
         do i=1,nx
             fmi=fm(1+(i-1)/3)
             an(i)=-dtt*d1(i)/fmi
             vn(i)=vn(i)+2.d0*an(i)
             ekin=ekin+fmi*vn(i)*vn(i)
         enddo
         ekin=0.5d0*ekin/(dt*dt)
         if(iter.eq.1) sekin=0.0
         sekin=sekin+ekin
         if(mod(iter,20).eq.1) then
             ekav=sekin/(20*nx)
             temp1=1006.8*ekav               ! 1006.8=2/Kb, Kb=0.00197 kcal/mol K .
             fact=0.08
             sekin=0.0
             if(temp.lt.temp1) fact=-0.08
             do i=1,nx                       ! Scales velocites to adjust temperature .
                 vn(i)=vn(i)*(1.0+fact)
             enddo
         endif
         return
         end
c========================================================================
         subroutine energy

c        Calculates the energy and forces in the system

         implicit real*8(a-h,o-z)
         parameter (ndip=500,natom=2*ndip,ncoord=3*natom)
         real*8 dr(3)
         common/cntrl/iter,niter,epot,ekin,etot,temp1
         common/coords/nd,ns,nsolv,nr,iac(20,4),x(3,natom),xs(3,20),
```

```
&                         nb(10),ib(10,4),jb(10,4)
          common/mapping/fl(4),emap(4,10000),imap
          common/force/du(3,natom),dx(3)
          common/param/crgs(20,4),crg(2),avdw(8),bvdw(8),fmass(8),
&                         rz(8),am(8),d(8),r12,dt,temp,rad
          data avdw/200.,200.,600.,600.,1000.,600.,600.,600./
          data bvdw/12.,24.,24.,24.,24.,24.,24.,24./
          data rz/.74,1.32,1.4,1.54,2.08,1.99,2.28,2.67/
          data am/1.99,2.32,2.3,1.60,1.83,2.03,1.94,1.85/
          data d/104.,54.,58.,88.,51.,58.,47.,36./
          epot=0.d0
          nsolv=2*nd
          ntot=nsolv+ns
          do i=1,ntot
             do k=1,3
                   du(k,i)=0.d0
             enddo
          enddo
          do 10 i=1,nsolv-2                              ! Calculates the solvent-solvent interactions.
             mi=1+mod(i-1,2)
             id=(i+1)/2
             do 20 j=i+1,nsolv
                   mj=1+mod(j-1,2)
                   jd=(j+1)/2
                   if(id.eq.jd) go to 20
                   r2=0.0
                   do k=1,3
                         dr(k)=x(k,i)-x(k,j)
                         r2=r2+dr(k)**2
                   enddo
                   r2=1./r2
                   r1=dsqrt(r2)
                   avd=avdw(mi)*avdw(mj)
                   bvd=bvdw(mi)*bvdw(mj)
                   r6=r2*r2*r2
                   cij=332.d0*crg(mi)*crg(mj)
                   fij=avd*r6**2-bvd*r6+cij*r1
                   epot=epot+fij
                   dfij=-r2*(12d0*avd*r6**2-6*bvd*r6+cij*r1)
                   do k=1,3
                         du(k,j)=du(k,j)-dfij*dr(k)
                         du(k,i)=du(k,i)+dfij*dr(k)
                   enddo
20            continue
10        continue
          fk=400.d0
          l=0
          do i=1,nd                                      ! The solvent intramolecular interactions.
             l=l+1
             do k=1,3
                   dr(k)=x(k,l)-x(k,l+1)
             enddo
             r2=dr(1)*dr(1)+dr(2)*dr(2)+dr(3)*dr(3)
```

```
              r1=dsqrt(r2)
              epot=epot+0.5d0*fk*(r1-r12)**2
              dfij=fk*(r1-r12)
              dfij=dfij/r1
              do k=1,3
                   du(k,l)=du(k,l)+dfij*dr(k)
                   du(k,l+1)=du(k,l+1)-dfij*dr(k)
              enddo
              l=l+1
        enddo
        imap=imap+1
        emap(1,imap)=0.
        emap(2,imap)=0.
        do 50 i=1,ns                                    ! The solute solvent interactions .
              do 50 j=1,nsolv
                   mj=1+mod(j-1,2)
                   r2=0.
                   do k=1,3
                        dr(k)=x(k,i+nsolv)-x(k,j)
                        r2=r2+dr(k)**2
                   enddo
                   r1=dsqrt(r2)
                   r1=1.d0/r1
                   r2=r1**2
                   r6=r2**3
                   do ii=1,nr
                        aij=avdw(iac(i,ii))*avdw(mj)
                        bij=bvdw(iac(i,ii))*bvdw(mj)
                        crgt=332*crgs(i,ii)*crg(mj)
                        fij=crgt*r1+aij*r6**2-bij*r6
                        emap(ii,imap)=emap(ii,imap)+fij     ! Stores the Evb energy of each state .
                        fij=fij*fl(ii)
                        epot=epot+fij
                        dfij=-r2*(12d0*aij*r6**2-6*bij*r6+crgt*r1)
                        dfij=dfij*fl(ii)
                        do k=1,3
                             du(k,j)=du(k,j)-dfij*dr(k)
                             du(k,i+nsolv)=du(k,i+nsolv)+dfij*dr(k)
                        enddo
                   enddo
50      continue
        call solute
        return
        end
c==============================================================
c      subroutine solute
c
c      This subroutine evaluates the solute intramolecular contributions
c      to the energy.

       implicit real*8(a-h,o-z)
       parameter (ndip=500,natom=2*ndip,ncoord=3*natom)
       common/cntrl/iter,niter,epot,ekin,etot,temp1
       common/coords/nd,ns,nsolv,nr,iac(20,4),x(3,natom),xs(3,20),
```

```
    &                    nb(10),ib(10,4),jb(10,4)
          common/dynmcs/fm(natom),vn(ncoord),anm1(ncoord),an(ncoord)
          common/force/du(3,natom),dx(3)
          common/mapping/fl(4),emap(4,10000),imap
          common/param/crgs(20,4),crg(2),avdw(8),bvdw(8),fmass(8),
    &                    rz(8),am(8),d(8),r12,dt,temp,rad
          do ii=1,nr
              do 10 i=1,ns-1
                  i1=i+1
                  do 10 j=i1,ns
                      nb1=nb(ii)
                      inon=1
                      do k=1,nb1
                          if(i.eq.ib(k,ii).and.j.eq.jb(k,ii)) inon=0
                      enddo
                      i1=i+nsolv
                      j1=j+nsolv
                      r2=0.
                      do k=1,3
                          dx(k)=x(k,i1)-x(k,j1)
                          r2=r2+dx(k)**2
                      enddo
                      r1=sqrt(r2)
                      if(mod(iter,80).eq.1.and.inon.eq.0) write(6,*) 'b=',i,j,r1
                      c=332*crgs(ii,i)*crgs(ii,j)
                      aij=avdw(iac(i,ii))*avdw(iac(j,ii))
                      bij=bvdw(iac(i,ii))*bvdw(iac(j,ii))
                      r0=0.5*(rz(iac(i,ii))+rz(iac(j,ii)))
                      fa=0.5*(am(iac(i,ii))+am(iac(j,ii)))
                      d1=dsqrt(d(iac(i,ii))*d(iac(j,ii)))
                      call fintra(f,df,r1,d1,fa,r0,c,aij,bij,inon)
                      epot=epot+f*fl(ii)
                      df=df*fl(ii)
                      emap(ii,imap)=emap(ii,imap)+f
                      do k=1,3
                          du(k,i1)=du(k,i1)+df*dx(k)/r1
                          du(k,j1)=du(k,j1)-df*dx(k)/r1
                      enddo
 10             continue
          enddo
          return
          end
c===========================================================================
          subroutine fintra(f,df,r1,d,fa,r0,c,a,b,inon)
          implicit real*8(a-h,o-z)
          if(inon.eq.0) then
              ex=exp(-fa*(r1-r0))
              f=d*(ex*ex-2.*ex)
              df=-2*fa*d*(ex*ex-ex)
              f=f+10*(r1-r0)**2          ! Quadratic constraint to keep bonded atoms
              df=df+20*(r1-r0)           ! within a reasonable distance.
          else
              r2=1/(r1**2)
              r6=r2**3
```

```
                    r12=r6*r6
                    f=a*r12-b*r6+c/r1
                    df=-12.*a*r12+6.*b*r6-c/r1
                    df=df/r1
              endif
              return
              end
c===================================================================
              subroutine grid

c             This subroutine generates the initial grid of solvent molecules .

              implicit real*8(a-h,o-z)
              parameter (ndip=500,natom=2*ndip,ncoord=3*natom)
              parameter (ngrid=8)
              real*8 dx
              real*8 xt(3,ngrid*ngrid*ngrid),xsum(3),xl(3)
              real*8 theta,phi,pi,e(3)
              common/coords/nd,ns,nsolv,nr,iac(20,4),x(3,natom),xs(3,20),
             &                    nb(10),ib(10,4),jb(10,4)
              common/param/crgs(20,4),crg(2),avdw(8),bvdw(8),fmass(8),
             &                    rz(8),am(8),d(8),r12,dt,temp,rad
              data xsum/3*0.d0/
              dx=3.0
              n=0
              do 30 i1=1,ngrid
                  do 30 i2=1,ngrid
                      do 30 i3=1,ngrid
                          n=n+1
                          xt(1,n)=i1*dx
                          xt(2,n)=i2*dx
                          xt(3,n)=i3*dx
                          do k=1,3
                                  xsum(k)=xsum(k)+xt(k,n)
                          enddo
30            continue
              do k=1,3
                  xsum(k)=xsum(k)/n
              enddo
              do i=1,n
                  do k=1,3
                          xt(k,i)=xt(k,i)-xsum(k)
                  enddo
              enddo
              rad2=rad*rad
              write(6,*) 'rad',rad,rad2
              nd=0
              nst=ns
              if(ns.eq.0) nst=1
              do i=1,n
                  do j=1,nst
                      rmin=1.d6
                      r2=0.
                      do k=1,3
                              xl(k)=xs(k,j)
```

```
                    if(ns.eq.0) xl(k)=0.
                    r2=r2+(xt(k,i)-xl(k))**2
              enddo
              if(rmin.gt.r2)rmin=r2
         enddo
         if (rmin.le.rad2.and.rmin.gt.5.0) then
              nd=nd+1
              do k=1,3
                    xt(k,nd)=xt(k,i)
              enddo
         endif
      enddo
1000  format (2x,3f10.3)
      pi=dacos(-1.d0)
      l=0
      do i=1,nd
         l=l+1
         theta=pi*randm(ir)
         phi=2.d0*pi*randm(ir)
         e(1)=dsin(theta)*dcos(phi)          ! e(1), e(2), e(3) are unit vectors.
         e(2)=dsin(theta)*dsin(phi)
         e(3)=dcos(theta)
         do k=1,3
              x(k,l)=xt(k,i)+0.5*r12*e(k)
              x(k,l+1)=xt(k,i)-0.5*r12*e(k)
         enddo
         l=l+1
      enddo
      return
      end
c===================================================================
      function randm(i)
      double precision a,p,x
      data a/16807.0d0/, p/2147483647.0d0/, c/4.6566129e-10/
      x=i
      x=a*x
      x=dmod(x,p)+.5d+0
      i=x
      randm=x
      randm=c*randm
      return
      end


c===================================================================
c===================================================================


Data for 3.A (evaluation of the radial distribution function)

1.1 .005 200. 6.0 0.3
+0.4
-0.4
0    ns    (for pure solvent calculations)
0    nr
0    nb
```

3.B. Calculations of the Free Energy Associated with Changing the Ion Charge from 0 to +1

Use program 3.A

c==
c==

Data for 3.B (the free energy calculations)

```
1.1 .0005 200. 6.0 0.3
+0.4
-0.4
1     ns
2     nr
0     nb
4     iac
0.0   crgs
0     nb
4     iac
+1.0  crgs
0.0 0.0 0.0
```

3.C. $S_N 2$ Free Energy Profile

Use program 3.A

c==
c==

Data for 3.C

```
1.1 .01 200. 6.0 0.3
+0.4
-0.4
3     ns
2     nr
1     nb
2 3   ib jb
6 4 6      iac
-1.0 0.0 0.0  crgs
1             nb
1 2          ib jb
6 4 6        iac
0.0 0.0 -1.0  crgs
0.0 0.0 0.0
3.0 0.0 0.0
4.8 0.0 0.0
```

REFERENCES

1. A. Warshel and S. T. Russell, *Quart. Rev. Biophys.*, **17**, 283 (1984).
2. O. Matsuoka, E. Clementi, and M. Yoshimine, *J. Chem. Phys.*, **64**, 1351 (1976).
3. S. Kuwajima and A. Warshel, *J. Phys. Chem.*, **94**, 460 (1990).
4. D. A. McQuarrie, *Statistical Mechanics*, Harper & Row, New York, 1976.
5. H. Gould and J. Tobochnik, *Computer Simulation Methods, Applications to Physical Systems*, Addison-Wesley Publishing Co., Reading, MA, 1988.
6. *Simulation of Liquids and Solids*, G. Ciccotti, D. Frenkel, and I. R. McDonald (Eds.), North-Holland Physics Publishing, Amsterdam, 1987.
7. A. Rahman and F. H. Stillinger, *J. Chem. Phys.*, **55**, 3336 (1971).
8. J. P. Valleau and G. M. Torrie, in *Modern Theoretical Chemistry* B. J. Berne, (Ed.), Vol. 5, Plenum, New York, 1977, pp. 169.
9. M. Born, *Z. Phys.*, **1**, 45 (1920).
10. G. King and A. Warshel, *J. Chem. Phys.*, **91**, 3647 (1989).
11. (a) J.-K. Hwang, G. King, S. Creighton, and A. Warshel, J. Am. Chem. Soc., **110**, 5297 (1988). (b) J.-K. Hwang, S. Creighton, G. King, D. Whitney, and A. Warshel, *J. Chem. Phys.*, **89**, 859 (1988).
12. C. H. Bennett in *Algorithms for Chemical Computations*, R. E. Christoferson, (Ed.), ACS, Washington D.C., 1977.
13. R. A. Marcus, *J. Chem. Phys.*, **72**, 891 (1968).
14. J. N. Brönsted and K. Pederson, *Z. Phys. Chem.*, **108**, 185 (1924).
15. G. S. Hammond, *J. Am. Chem. Soc.*, **77**, 334 (1955).
16. J. W. Albery and M. M. Kreevoy, *Adv. Phys. Org. Chem.*, **16**, 87 (1978).
17. S.S. Shaik, *Prog. Phys. Org. Chem.*, **15**, 197 (985).
18. B. J. Gertner, K. R. Wilson, and J. T. Hynes, *J. Chem. Phys.*, **90**, 3537 (1989).
19. D. Chandler, *Chem. Scr.*, **29A**, 61 (1989).

4

POTENTIAL SURFACES AND SIMULATIONS OF MACROMOLECULES

4.1. BACKGROUND

Macromolecules are formed from many fragments of smaller molecules which are connected to each other by covalent bonds. For example, protein molecules are assembled from *amino acids* which are interconnected by peptide bonds (see Fig. 4.1). Typical amino acids are given in Fig. 4.2.

A folded protein might look like an extremely complicated object in comparison to the small molecules considered in the previous chapters. Yet, the fragments of proteins and other macromolecules are assembled by the same type of bonds that connect the atoms in small molecules. This fact can be exploited in constructing potential surfaces for macromolecules and representing them in a way not much different than that used in Chapter 3 for solvent molecules. That is, we will see below that a macromolecule can be described as a collection of balls connected by springs, while interacting with nonnearest neighbors through charge–charge Coulomb interactions and short-range repulsions.

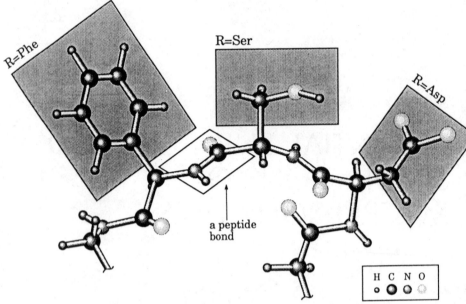

FIGURE 4.1. A protein is assembled from amino acids connected to each other by peptide bonds. Each amino acid contributes an identical group to the backbone plus a distinguishing residue (R) as a side chain.

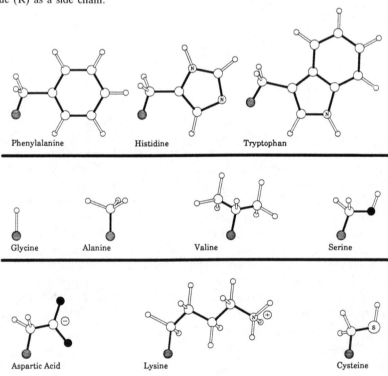

FIGURE 4.2. Some commonly occurring amino acids.

4.2. FORCE FIELDS FOR LARGE MOLECULES

4.2.1. The Forces in Macromolecules Can Be Described by Simple Functions

A reliable evaluation of the Born–Oppenheimer potential surfaces of large molecules by quantum mechanical treatments is impractical even with the current generation of computers. Alternatively, one can look for a simple and yet reliable empirical approximation for the Born–Oppenheimer surface of polyatomic molecules. At first sight it might seem hopeless to approximate an unknown multidimensional surface by some empirical function. However, the molecular potential surface is not really so complicated. Apparently, the variation of the potential as a function of bond stretching or bond angle bending depends mainly on the neighboring atoms. Thus, if one deals with a long-chain molecule, the proper representation of the potential is not by an arbitrary function of the Cartesian coordinates, but rather by a sum of contributions from subspaces of bonded atoms and contributions from interactions between nonbonded atoms (which represents the coupling between the subspaces). In general, it is possible to write the potential surface as a sum of short-range interactions between atoms bonded to each other or to the same atom (1–2 and 1–3 interactions), intermediate interactions between atoms separated by two atoms (1–4 interactions) and nonbonded interactions between other atoms. Such a potential can be given in the following form:

$$U(\mathbf{s}) = U_{b,\theta}(\mathbf{b}, \boldsymbol{\theta}) + U_{\phi}(\boldsymbol{\phi}) + U(\mathbf{r}) \tag{4.1}$$

where $\mathbf{s}$ is the vector of internal coordinates composed of $\mathbf{b}$, $\boldsymbol{\theta}$, $\boldsymbol{\phi}$, and $\mathbf{r}$, which are, respectively, the vectors of bond lengths, bond angles, torsional angles, and nonbonded distances. The first three terms define a very deep potential well, and since the molecule stays in most cases inside this well (except in the extreme case of bond dissociation), it is reasonable to approximate this part of the potential surface by its quadratic expansion, which is given by

$$U_{b\theta}(\mathbf{b}, \boldsymbol{\theta}) = \frac{1}{2} \sum_i K_{b,i}(b_i - b_{0,i})^2 + \frac{1}{2} \sum_i K_{\theta,i}(\theta_i - \theta_{0,i})^2 + \text{cross terms} \tag{4.2}$$

where U is usually given in kcal/mol, b in Å and θ in radians. The torsional potential $U(\phi)$ is a periodic function, which can be described by the leading terms in the Fourier expansion of the potential

$$U_{\phi}(\boldsymbol{\phi}) = \frac{1}{2} \sum_i K_{\phi,i}(1 - \cos n\phi_i) \tag{4.3}$$

The nonbonded potential can be described by an atom–atom interaction potential of the form used in eqs. (3.1) and (3.2).

$$U_{\mathrm{nb}} = \sum_{ij} A_{ij} r_{ij}^{-12} - B_{ij} r_{ij}^{-6} + 332 q_i q_j / r_{ij} + U_{\mathrm{ind}}(r) \tag{4.4}$$

where U_{nb} is given in kcal/mol, r_{ij} in Å, and q in au. Potential surfaces of the form used in eq. (4.1) are frequently called "force fields" (see Refs. 1 and 2).

4.2.2. Refining the parameters in Molecular Force Fields

The parameters in molecular force fields (e.g., K_b, K_θ, A, and B) can be determined by using them to calculate different independent molecular

TABLE 4.1. Some Simplified CFF Parameters for Hydrocarbons and Amides[a]

Bond-Stretching Parameters			Angle-Bending Parameters				
Bond	$\frac{1}{2}K_b$	b_0	Angles	$\frac{1}{2}K_\theta$	θ_0	$\frac{1}{2}F_{ik}$[b]	r_{ik}^0
CH	286.0	1.099	HCH	40.0	1.911	2	1.8
CC	110.0	1.490	CCH	25.0	1.911	43	2.2
C'N	403.0	1.278	CCC	18.0	1.911	55	2.5
C'O	595.0	1.200	C'NH	26.0	2.094	27	2.0
CN	201.0	1.457	C'NC	54.0	2.094	10	2.4
NH	405.0	0.980	HC'O	48.0	2.094	90	2.3
			NC'C	41.0	2.094	52	2.4

Torsional Parameters			Nonbonded Parameters[c]				
Angle	$\frac{1}{2}K_\phi$	n	Nonbonded	ε	r^*	q	α
CCCC	1.2	3	H	0.05	3.4	0.1	0.1
CC'NC	12.0	2	C	0.42	3.6	−0.1	1.2
C'NCC'	−1.5	3	C'	0.40	3.6	0.5	1.2
NCC'N	0.5	3	N	0.36	3.6	−0.4	1.2
			O	0.42	3.2	−0.5	1.2
			H(N)	0.10	0.4	0.3	0.0

[a]Energies in kcal/mol, lengths in Å, angles in radians and q in au. C' and H(N) are, respectively, the carbonyl carbon and hydrogen of the amide bond. See Ref. 1 for more details.
[b]Cross terms for the angle terms are introduced by adding the Urey-Bradley term $\frac{1}{2}F_{ik}(r_{ik} - r_{ik}^0)^2$, where the given angle is formed by the atoms i, j and k.
[c]The parameters A and B are defined here by $A_{ij} = \varepsilon_{ij}(r_{ij}^*)^{12}$, $B_{ij} = 2\varepsilon_{ij}(r_{ij}^*)^6$ where $\varepsilon_{ij} = (\varepsilon_i \varepsilon_j)^{1/2}$ and $r_{ij}^* = (r_i^* + r_j^*)/2$.

properties (e.g., energies, structures, and vibrations) and then fitting the calculated properties to the corresponding observed properties by a systematic change of the potential parameters in a least-squares procedure. In general the parameters b_0 and θ_0 are sensitive to structural information and the parameters K_b and K_θ are sensitive to molecular vibrations, while the K_ϕ are determined by information about torsional barriers (that can also be obtained effectively by quantum mechanical calculations of such barriers). The parameters A and B of the nonbonded potential are obtained by fitting properties of molecular crystals (e.g., sublimation energies, crystal structures, and lattice vibrations). Nevertheless, all properties depend in one way or another on all parameters; thus it is important to perform the fitting procedure simultaneously for different properties. Such a simultaneous fitting is the basis of the so-called *consistent force field* (CFF) approach (Ref. 1) (for related widely used methods see for example Ref. 2). A typical parameter set is given in Table 4.1.

4.3.   ENERGY MINIMIZATION

4.3.1.   The Steepest Descent Method

With analytical potential functions we can try to evaluate the molecular equilibrium geometries and the vibrations around these configurations. This task can be accomplished in the simplest way using the Cartesian representation (Ref. 1.) That is, the potential surface for a molecule with n atoms can be expanded formally around the equilibrium configuration $\mathbf{r}_0$ and give

$$U(\mathbf{r}_0 + \delta\mathbf{r}) = U(r_0) + \sum_{i\alpha} (\partial \overset{\ast}{U}/\partial r_{i\alpha})\delta r_{i\alpha}$$

$$+ \frac{1}{2} \sum_{i\alpha, j\beta} (\partial^2 U/\partial r_{i\alpha}\, \partial r_{j\beta})\delta r_{i\alpha} \delta r_{j\beta} + \cdots \qquad (4.5)$$

where the indices i and j designate atoms while α and β run over the x, y, and z coordinates of each atom. The first term in eq. (4.5) is just the energy of the system at equilibrium. The second term represents the deviation from equilibrium and the $3n$ set of equations (for $i = 1, 2, \ldots n$ and $\alpha = x, y, z$)

$$\partial U/\partial r_{i\alpha} = 0 \qquad (4.6)$$

represents the condition that $\mathbf{r}_0$ is an equilibrium configuration. This set of equations can be solved approximately by the steepest descent method, where the nth step toward the minimum is obtained by

$$r_{i\alpha}^n = r_{i\alpha}^{n-1} - g\, \frac{(\partial U/\partial r_{i\alpha})}{|\Sigma_{j\beta}\, (\partial U/\partial r_{j\beta})^2|^{1/2}} \qquad (4.7)$$

here g is a scaling factor which is changed in an iterative way to prevent $\mathbf{r}$ from overshooting the minimum [g is increased in every iteration as long as $U(\mathbf{r}^n) < U(\mathbf{r}^{n-1})$ and decreased if $U(\mathbf{r}^n) > U(\mathbf{r}^{n-1})$]. This approach is demonstrated in Program 4.A.

Exercise 4.1. Use the $\frac{1}{2} K_b (b - b_0)^2$ term in the potential of eq. (4.2) (with $K_b = 100$ and $b_0 = 1.5$ Å) as the molecular potential for a ring of three carbon atoms and find an equilibrium geometry using eq. (4.7) and Program 4.A.

Solution 4.1. The potential for our system can be written as $U = 50\{(b_{12} - 1.5)^2 + (b_{23} - 1.5)^2 + (b_{31} - 1.5)^2\}$. Next we should express the bond lengths b and the derivatives of U in terms of the Cartesian coordinates of the system. This is done with $b_{ij}^2 = \{\Sigma_{\alpha=1}^3 (r_{i\alpha} - r_{j\alpha})^2\}$ and $\partial U/\partial r_{i\alpha} = \Sigma_{ij} (\partial U/\partial b_{ij})(\partial b_{ij}/\partial r_{i\alpha})$ with $\partial U/\partial b_{ij} = K(b_{ij} - b_0)$ and $\partial b_{ij}/\partial r_{i\alpha} = (r_{i\alpha} - r_{j\alpha})/b_{ij}$. Now we can start with an initial guess of coordinates $\mathbf{r}^{(1)}$ and then use eq. (4.7) and Program 4.A to change $\mathbf{r}$ and minimize $U(\mathbf{r})$.

4.3.2. Converging Minimization Methods

The steepest-descent method converges very slowly and is not very effective in searching for minima. A much more reliable and efficient approach is the modified Newton–Raphson method. This method is based on expanding the gradient as a Taylor series around the given $\mathbf{r}$ and finding the $\delta\mathbf{r}$ that leads to $\mathbf{r}_0$ where the gradient is zero (i.e. $\mathbf{r}_0 = \mathbf{r} + \delta\mathbf{r}$). This gives

$$\partial U(\mathbf{r}_0)/\partial r_{i\alpha} = \partial U(\mathbf{r}_0 + \delta\mathbf{r})/\partial r_{i\alpha} + \sum_{j\beta} F_{i\alpha,j\beta}\, \delta r_{j\beta} = 0 \qquad (4.8)$$

where $\mathbf{F}$ is the matrix of second derivatives, i.e., $F_{i\alpha,j\beta} = \partial^2 U/\partial r_{i\alpha} \partial r_{j\beta}$. Then, solving eq. (4.8), one obtains

$$\mathbf{r}_0 = \mathbf{r} + \delta\mathbf{r} = \mathbf{r} - \mathbf{F}^+ \nabla U(r) \qquad (4.9)$$

where ∇U is the gradient vector [with $(\nabla U)_{i\alpha} = \partial U/\partial r_{i\alpha}$] and $\mathbf{F}^+$ is the generalized inverse of $\mathbf{F}$, which is constructed by "filtering" the zero eigenvalues of $\mathbf{F}$ before inverting this matrix. This is done by diagonalizing $\mathbf{F}$ with $\boldsymbol{\Lambda} = \mathbf{S}'\mathbf{F}\mathbf{S}$, constructing the matrix $\bar{\Lambda}$ with $\bar{\Lambda}_{ii} = \Lambda_{ii}$ for $\Lambda_{ii} \neq 0$ and $\bar{\Lambda}_{ii} = \infty$ for $\Lambda_{ii} = 0$ and then obtaining $\mathbf{F}^+$ by $\mathbf{F}^+ = \mathbf{S}(\bar{\Lambda}^{-1})\mathbf{S}'$. The reason for introducing this special inversion procedure is that $U(r)$ is invariant to rigid rotation and translation of the molecule so that the $\mathbf{F}$ matrix has zero eigenvectors which prevent its regular inversion (Ref. 1).

It is instructive to note that both the steepest-descent and the Newton–Raphson methods lead in the direction of $-\nabla U$; however, the steepest-descent method is unable to tell us how far to go in each step and therefore we have to search for the minimum in a very ineffective way (see Fig. 4.3).

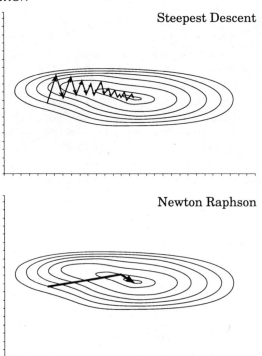

Steepest Descent

Newton Raphson

FIGURE 4.3. Illustrating the effectiveness of different minimization schemes. The steepest-descent method requires many steps to reach the minimum, while the Newton–Raphson method locates the minimum in a few steps (at the expense, however, of evaluating the second derivative matrix).

Exercise 4.2. Use the Newton–Raphson method to minimize the function $u = (x - 1)^2 + (y - 2)^2$ (Hint: in the present case $\mathbf{F}^+$ is simply the inverse of $\mathbf{F}$) so that $\mathbf{F}^{-1}\mathbf{F} = \mathbf{F}^+\mathbf{F} = \mathbf{I}$.

Solution 4.2. The matrix $\mathbf{F}$ is given by $F_{11} = (\partial^2 U/\partial x^2) = 2$, $F_{22} = 2$, $F_{12} = (\partial^2 U/\partial x \partial y) = 0$, $F_{21} = F_{12}$. The inverse of $\mathbf{F}$ is given by $(\mathbf{F}^{-1})_{11} = (\mathbf{F}^{-1})_{22} = \frac{1}{2}$; $(\mathbf{F}^{-1})_{12} = (\mathbf{F}^{-1})_{21} = 0$ (this can be easily verified by evaluating $\mathbf{F}^{-1}\mathbf{F}$ and seeing that the product is indeed the unit matrix). Now we just use eq. (4.9), getting $x = x - F_{11}^{-1}(\partial U/\partial x) = x - (1/2) \cdot (2(x - 1))$; $y = y - F_{22}^{-1}(\partial U/\partial y) = y - (1/2) \cdot (2(y - 2))$.

Exercise 4.3. Use the Newton–Raphson method to minimize the system in Problem 4.1.

Solution 4.3. Use Program 4.B.

Equation (4.9) requires the evaluation of the second derivative matrix $\mathbf{F}$, which is quite involved. Alternatively, one can use the *conjugated gradient* methods where an approximation of $\mathbf{F}^+$ is being built while searching the minimum using only the first derivatives vector ∇U (for a description of these powerful methods and related approaches, see Ref. 3).

Energy minimization methods that exploit information about the second derivative of the potential are quite effective in the structural refinement of proteins. That is, in the process of X-ray structural determination one sometimes obtains bad steric interactions that can easily be relaxed by a small number of energy minimization cycles. The type of relaxation that can be obtained by energy minimization procedures is illustrated in Fig. 4.4. In fact, one can combine the potential $U(\mathbf{r})$ with the function which is usually optimized in X-ray structure determination (the "R factor") and minimize the sum of these functions (Ref. 4) by a conjugated gradient method, thus satisfying both the X-ray electron density constraints and steric constraint dictated by the molecular potential surface.

Although the conjugated gradient and related methods are very effective in finding *local minima*, they do not overcome the problems associated with the enormous dimensionality of macromolecules. That is, in systems with

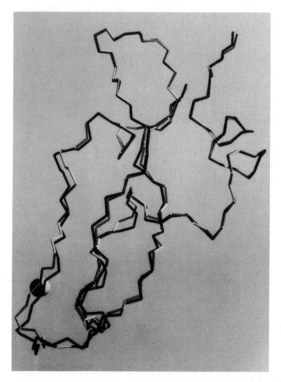

FIGURE 4.4. The relaxation of steric interactions by energy minimization. The figure shows the result of an energy minimization of the Gly 12 → Ala mutant of BPTI from the native geometry (light) where it has bad steric contacts, to a relaxed geometry (dark). The mutant Alanine residue is drawn as a sphere.

many degrees of freedom we expect to find a very large number of local minima and it is not clear, for example, how to find in an efficient way the lowest minimum. For this purpose one must use much more computer time and different types of search procedures (Monte Carlo or molecular dynamics), generating different configurations and minimizing at different regions of the conformation space of the given molecule. Such methods were, however, useless until the emergence of supercomputers in the early 1980s and even at the present time they do not allow for a complete configuration search in proteins.

4.4. NORMAL MODES ANALYSIS OF LARGE MOLECULES

Finding a local minimum by a convergent minimization method allows one to exploit the third term in eq. (4.5) for evaluation of the vibrations around that minimum. That is, the potential for infinitesimal vibrations around the local minimum $\mathbf{r}_0$ can be written

$$\delta U(\mathbf{r}_0) = \frac{1}{2}\, \delta\mathbf{r}'\mathbf{F}(\mathbf{r}_0)\, \delta\mathbf{r} \tag{4.10}$$

The kinetic energy for the system can be written as

$$\delta T = \frac{1}{2}\, \delta\dot{\mathbf{r}}'\mathbf{M}\, \delta\dot{\mathbf{r}} = \frac{1}{2} \sum_{i,\alpha} m_i\, \delta\dot{r}_{i,\alpha}^2 \tag{4.11}$$

where $\mathbf{M}$ is the diagonal matrix of the atomic masses ($M_{i\alpha,j\beta} = M_i \delta_{ij} \delta_{\alpha\beta}$). The expression for the potential energy can be greatly simplified if we transform the Cartesian coordinates to a new set of coordinates called *normal coordinates* using

$$\delta\mathbf{r} = \mathbf{M}^{-1/2}\mathbf{L}\mathbf{Q} \tag{4.12}$$

where the matrix $\mathbf{L}$ is constructed from the column vectors $\mathbf{L}^s$ that are obtained by diagonalizing the mass weighted $\mathbf{F}$ matrix

$$(\mathbf{M}^{-1/2}\mathbf{F}\mathbf{M}^{-1/2})\mathbf{L}^s = \lambda_s \mathbf{L}^s \tag{4.13}$$

With this transformation matrix we can express the kinetic and potential energies in the simple form

$$\delta U = \frac{1}{2} \sum_s \lambda_s Q_s^2$$

$$\delta T = \frac{1}{2} \sum_s \dot{Q}_s^2 \tag{4.14}$$

Using the Lagrange's equation of motion (Ref. 6) (the Lagrange's equation $\{d[\partial \, \delta T/\partial \dot{Q}_s]/dt + \partial \, \delta U/\partial Q_s\} = 0$ may be considered as an alternative form of Newton's equation of motion) and $Q_s(t) = A_s \cos(2\pi\nu_s t)$ gives

$$\lambda_s = (2\pi\nu_s)^2 \tag{4.15}$$

where ν_s is now identified as the vibrational frequency associated with the normal mode Q_s. Thus we obtain from eqs. (4.12) and (4.13) the vibrations of the molecule in terms of the $\delta\mathbf{r}_s$ vector.

The treatment described above (which was introduced in Ref. 1) is much simpler than the standard treatment (which uses internal coordinates b, θ, ϕ) (Ref. 5) and it can be conveniently implemented in studies of large molecules or small proteins, evaluating the second derivative matrix $\mathbf{F}$ numerically, using analytical first derivatives.

Exercise 4.4. Take the minimized structure of the three-carbon ring of Exercise 4.1 and evaluate its normal modes.

Solution 4.4. At the minimized structure of the ring ($\mathbf{r} = \mathbf{r}_0$) you have to obtain the matrix $\mathbf{F}$. This can be accomplished by numerical derivative of the vector ∇U, evaluated analytically in Exercise 4.1. The expression to be used is

$$\mathbf{F}_{i\alpha} = \lim_{\varepsilon \to 0} (\nabla U(\mathbf{r}_0 + \varepsilon \, \delta\mathbf{r}_{i\alpha}) - \nabla U(\mathbf{r}_0))/\varepsilon \tag{4.16}$$

where $\delta\mathbf{r}_{i\alpha}$ is a vector whose elements are set to zero except the $i\alpha$th element which is taken as 1. The parameter ε should have a finite value larger than the numerical precision of the given computer (typically, 10^{-6} Å). Once you construct the matrix $\mathbf{F}$ just diagonalize the matrix $\mathbf{M}^{-1/2}\mathbf{F}\mathbf{M}^{-1/2}$ and obtain the frequencies ν_s from the eigenvalues λ_s (by eq. 4.15) and the normal mode vectors which are simply the eigenvectors $\mathbf{L}_s$. The corresponding Cartesian displacement for each mode is obtained from eq. (4.12). You will obtain three modes with non-zero frequencies that correspond to intramolecular vibrations, and six with zero frequencies that correspond to translation and rotation of the molecule (for these modes the change in potential energy is zero and therefore $\lambda_s = 0$). The three intramolecular normal modes obtained from our exercise are shown in Fig. 4.5 (see Program 4.C, for a hint).

The harmonic normal mode description is quite useful for approximated evaluation of various molecular properties. For example, one can use this description in a convenient way to evaluate the average thermal atomic motion. This is done by using the normal mode vector $\mathbf{L}_s$ in eq. (4.12), which can be written as

$$\Delta r_{i\alpha} = m_i^{-1/2} \sum_s^{3n} L_{i\alpha}^s Q_s \tag{4.17}$$

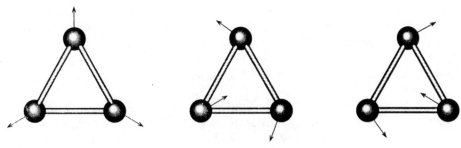

FIGURE 4.5. The normal modes of the CCC ring.

From this expression we obtain the thermal average

$$\langle \Delta r_{i\alpha} \Delta r_{i\alpha} \rangle_T = m_i^{-1} \sum_s (L_{i\alpha}^s)^2 \langle Q_s^2 \rangle_T$$

$$\langle Q_s^2 \rangle_T = (h/(4\pi^2 c \bar{\nu}_s)) \left[\frac{1}{2} + (\exp\{h\nu_s/k_B T\} - 1)^{-1} \right] \qquad (4.18)$$

where $\bar{\nu}$ is the frequency expressed in cm^{-1}.

The harmonic approximation can also be used to provide an estimate of the vibrational free energy, using (Refs. 1 and 6).

$$\Delta G_{\text{vib}} = \sum_s h\nu_s/2 + k_B T \ln[1 - \exp(-h\nu_s/k_B T)] \qquad (4.19)$$

This estimate should, however, be used while keeping in mind the availability of many minima in macromolecules and the fact that the ΔG_{vib} should in principle be averaged over these many minima. Furthermore, the effect of free energy associated with the solvent molecules is not included in ΔG_{vib}.

4.5. MOLECULAR DYNAMICS AND PHASE SPACE EXPLORATION

4.5.1. Thermal Amplitudes

With the powerful computers, that are now starting to emerge, one can use MD in the same way as outlined in Chapter 3 and explore the configuration space of large molecules. Using the potential functions of Section 4.2 and the procedure of Section 3.2.2, we can propagate trajectories of proteins at a given constant temperature. The results of such trajectories can be used to evaluate various average properties. For example, the average thermal atomic amplitudes can be expressed through eq. (3.4) as

$$\langle r_{i\alpha} r_{i\beta} \rangle = \int r_{i\alpha} r_{i\beta} \exp\{-U(\mathbf{r})\beta\} \, d\mathbf{r} \Big/ \int \exp\{-U(\mathbf{r})\beta\} \, d\mathbf{r} = \langle r_{i\alpha} r_{i\beta} \rangle_{\text{time}}$$

$$(4.20)$$

where we use the ergodic hypothesis to replace the Boltzmann average over the coordinates by time average over a long trajectory. Typical results for BPTI are given in Ref. 7.

4.5.2. Diffusion Constants and Autocorrelation Times

Various ligands "bind" to their protein sites in a diffusive motion. Similarly, the distance between different ends of a folded macromolecule changes in a way which can be described as a diffusive motion in the presence of a constraint potential (that keeps the parts of the molecule near their folded configurations). Brownian-type diffusive motion in the absence of a restrictive potential is characterized by a diffusion constant (Ref. 6)

$$D = \lim_{t \to \infty} \frac{1}{6t} \left\langle |\mathbf{r}(t) - \mathbf{r}(0)|^2 \right\rangle \tag{4.21}$$

The diffusion constant can be evaluated by a microscopic simulation using the relationship (Ref. 6)

$$D = \frac{1}{3} \int_0^\infty \left\langle \dot{\mathbf{r}}(0) \cdot \dot{\mathbf{r}}(t) \right\rangle \, dt \tag{4.22}$$

where the function $\left\langle \dot{\mathbf{r}}(0) \cdot (\dot{\mathbf{r}})(t) \right\rangle$ is called the *velocity autocorrelation function*. The diffusion constant is related to the corresponding friction constant γ by the *Einstein equation*

$$D = k_B T / \gamma \tag{4.23}$$

In order to evaluate the autocorrelation function we again exploit the ergodic hypothesis and replace the average over phase space $\langle \ \rangle$ by a time average writing,

$$C(t') = \left\langle \dot{\mathbf{r}}(0)\dot{\mathbf{r}}(t') \right\rangle = \frac{1}{\tau} \int_0^\tau \dot{\mathbf{r}}(t)\dot{\mathbf{r}}(t + t') \, dt \tag{4.24}$$

In calculating eq. (4.24) we divide the time axis to N equal intervals, evaluating

$$C(m) = \frac{1}{N} \sum_{i=1}^N \dot{\mathbf{r}}(i)\dot{\mathbf{r}}(i + m) \tag{4.25}$$

where the time is written here in integers. In order to understand this equation it is best to try it on a specific example such as that of Fig. 4.6. For example, in order to evaluate $C(t = 2)$ from Fig. 4.6 we have to collect the product $(\dot{x}(1)\dot{x}(3) + \dot{x}(2)\dot{x}(4) + \dot{x}(3)\dot{x}(5) + \cdots)$. The value of $C(t)$ changes from $\langle \dot{x}^2 \rangle$ for $\{C(0) = \frac{1}{N}[\dot{x}(1)\dot{x}(1) + \dot{x}(2)\dot{x}(2) + \cdots]\}$ to zero for large t

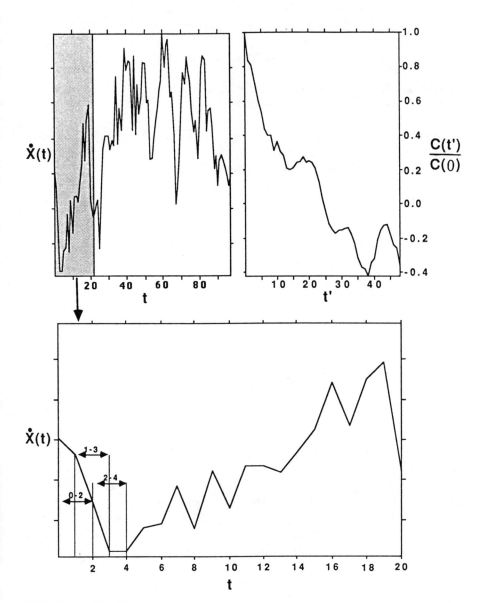

FIGURE 4.6. This illustration shows how to use the time dependance of $\dot{x}$ to construct an autocorrelation function $C(t) = \langle \dot{x}(0)\dot{x}(t) \rangle$. The upper part of the figure shows $\dot{x}(t)$ and the corresponding normalized $C(t)$. The lower part of the figure shows the beginning portion of $\dot{x}(t)$ on an expanded scale and illustrates the construction of $C(2)$ by the sum of the products $\dot{x}(t)\dot{x}(t+2)$ (see text for further discussion).

since in this case the values of $\dot{x}$ are *uncorrelated* and their product will average to zero. The time it takes for $C(t)$ to drop to $(1/e)$ from its value at $t = 0$ is called the *autocorrelation time*. This is the time after which the given variable (i.e., $\dot{r}$) starts to lose its memory about its initial condition.

Exercise 4.5. Evaluate the autocorrelation function and the autocorrelation time for the function $\exp(-t)\cos(2t)$.

Macromolecular fluctuations are characterized by correlation times which are closely related to the underlined relaxation processes. Such relaxation times can be evaluated by classical trajectory simulations using

$$\tau_A^{-1} = \frac{\int \langle A(0)A(t)\rangle \, dt}{\langle A(0)A(0)\rangle} = \int_0^\infty \exp\{-t/\tau_A\} \, dt \tag{4.26}$$

where A is the property under consideration (which is taken relative to its average value) and it is assumed that the autocorrelation function can be characterized by a single exponential time-dependent behavior.

From the various autocorrelation times which characterized macromolecular fluctuations, those associated with the fluctuation of the electrostatic field from the protein on its reacting fragments are probably the most important (see Ref. 8). These autocorrelation times define the *dielectric relaxation times* for different protein sites and can be used to estimate dynamical effects on biological reactions (see Chapter 9 for more details).

4.5.3. Free Energies of Macromolecules

Probably the most important aspects of macromolecular function are not associated with true dynamical properties but with activation-free energies and binding free energies. Thus the most important feature of computer-intensive phase-space exploration by Monte Carlo or molecular dynamics methods is the ability to evaluate free energies by the free-energy perturbation (FEP) methods described in Chapter 3. Such free-energy calculations are rather meaningless without taking into account the solvent around the protein. Effective ways of including the solvent in free-energy calculation of solvated proteins will be described in the next section. Early FEP studies of the energetics of proteins are described in Refs. 10–13. More recent studies are reviewed in Ref. 14 and different applications are considered in subsequent chapters.

4.6. ELECTROSTATIC FREE ENERGIES AND DIELECTRIC EFFECTS IN MACROMOLECULES

Proteins do not exist as isolated entities in the gas phase. They are always solvated by water molecules or embedded in membranes. Trying to obtain a realistic description of solvated proteins requires one to handle the effect of the surrounding environment. This is particularly true when one deals with ionized groups in proteins, since the interactions between such groups are largely screened by the surrounding solvent. For example, the electrostatic interaction between two ionized groups is given through eq. (4.4) as

$$U_{nb} = 332 q_1 q_2 / r_{12} \tag{4.27}$$

This expression gives an energy of 100 kcal/mol and an enormous force $(\partial U / \partial r)$ of about 30 kcal mol^{-1} Å^{-1} for two ionized groups of identical charges which are held 3 Å apart. This would "break" a protein apart in the absence of other compensating forces. However, other polar groups of the proteins and/or water molecules are always presented around ionized groups. This polar surrounding provides a compensating force that can be represented by

$$\bar{U}_{nb} = 332 q_1 q_2 / (r_{12} d_{12}) \tag{4.28}$$

where d_{12} is considered a "dielectric constant" and is typically larger than 30, even inside a protein (Ref. 8). With such a large d_{12} the effective force between charged groups becomes relatively small (1 kcal mol^{-1} Å^{-1} for two charges 3 Å apart). Unfortunately, there is no general macroscopic prescription for the evaluation of the specific value of d_{12} as it represents the effect of the surrounding of the charges, which is different in different cases. Thus it is crucial to take into account the effect of the solvent around the protein in any realistic simulation study.

Including the solvent around a protein can be done, in principle, by the explicit approach of Chapter 3. Such a treatment, however, is very expensive, in view of the large number of water molecules needed to properly solvate the entire protein. Thus we will consider below two alternative approaches which allow one to effectively represent the solvent. The discussion of these models will be focused on electrostatic aspects where the importance of solvent effects is easily demonstrated.

4.6.1. The Protein Dipoles–Langevin Dipoles (PDLD) Model

The calculations of electrostatic energies in proteins can be formulated using macroscopic models (Ref. 9), but such models require knowledge of the protein "dielectric constant," which cannot be determined from macroscopic considerations; this "constant" is different in different parts of the protein (see Ref. 14 for discussion). Fortunately, the use of microscopic approaches allows one to avoid the issue of the dielectric constant altogether by explicitly including all the key electrostatic contributions. A simple and effective way of doing so is provided by the protein dipoles–Langevin dipoles (PDLD) model (Ref. 8). This model (which is illustrated in Fig. 4.7) considers a charge or charges in a reference region (region I) and evaluates

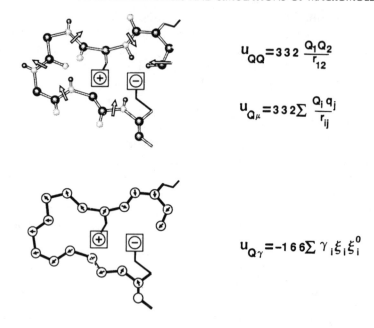

$$u_{QQ} = 332 \frac{Q_1 Q_2}{r_{12}}$$

$$u_{Q\mu} = 332 \sum \frac{Q_i q_j}{r_{ij}}$$

$$u_{Q\gamma} = -166 \sum \gamma_i \xi_i \xi_i^0$$

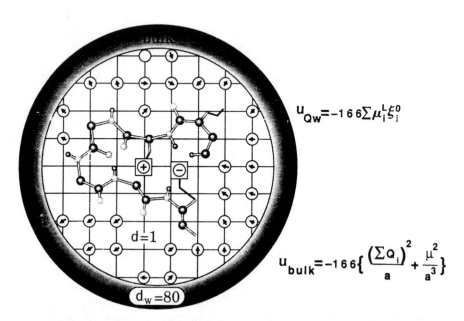

$$u_{Qw} = -166 \sum \mu_i^L \xi_i^0$$

$$u_{bulk} = -166 \left\{ \frac{\left(\sum Q_i\right)^2}{a} + \frac{\mu^2}{a^3} \right\}$$

FIGURE 4.7. A schematic description of the different contributions to the PDLD model. The figure considers the energetics of an ion pair inside a protein interior. The upper part describes the protein permanent dipoles, the middle part describes the induced dipoles of the protein, while the lower part describes the surrounding water molecules and the bulk region, which is represented by a macroscopic continuum model.

their interaction with the rest of the protein–solvent system. The protein atoms around region I (referred to as region II) are described by considering explicitly their residual charges and the induced dipoles associated with their atomic polarizabilities. The solvent molecules (region III) are described by the Langevin dipoles (LD) model of Chapter 2. The electrostatic energy of the charges in region I is evaluated by

$$U(Q) = U_{QQ} + U_{Q\mu} + U_{Q\gamma} + \Delta G_{Qw} \qquad (4.29)$$

where U_{QQ} is the interaction between the charges in region I and is given (in kcal/mol) by

$$U_{QQ} = 332 \sum_{i>j} Q_i Q_j / r_{ij} \qquad (4.30)$$

Note that we do not use any dielectric constant here since all interactions are considered explicitly. $U_{Q\mu}$ is the interaction between the charges in region I and the protein residual charges in region II and is given by

$$U_{Q\mu} = 332 \sum_{ij} Q_i q_j / r_{ij} \qquad (4.31)$$

$U_{Q\gamma}$ is the energy associated with polarizing the induced dipoles of the protein atoms and is given by [see eq. (3.3) and Ref. 8b]

$$U_{Q\gamma} = -166 \sum_i \gamma_i \xi_i \xi_i^0 \qquad (4.32)$$

where, ξ_i is the field on the ith atom. This field is evaluated in a self-consistent way including the field from the permanent charge of the system, ξ^0, and the induced dipoles ($\mu_i = \gamma_i \xi_i$) of the protein atoms. Finally, ΔG_{Qw} is the energy of polarization of the Langevin dipoles which represents the solvation of the system of the surrounding solvent and is given by [see eq. (2.20)]

$$\Delta Q_{Qw} = -166 \sum_i \mu_i^L \xi_i^0 \qquad (4.33)$$

where the μ_i^L are the Langevin dipoles around the protein which are polarized according to eq. (2.19) and reflects both the field ξ^0 from the permanent charges as well as the field ξ_μ from the other Langevin dipoles (Ref. 8). The contribution should also include the energy of the surrounding bulk region, which is evaluated by a macroscopic continuum model.

Exercise 4.2. Use the PDLD Model to calculate the energy of a charged Asp 3 in the Protein BPTI, considering only $U_{Q\gamma}$ and ΔG_{Qw}.

A reasonable alternative to the PDLD method can be obtained by approaches that represent the solvent as a dielectric continuum and evaluate the electric field in the system by discretized continuum approaches (see Ref. 15). Note, however, that the early macroscopic studies (including the pioneering studies of Ref. 9) have considered the protein as a nonpolar medium and could not account consistently for the energetics of charged groups (see Ref. 14 for discussion).

4.6.2. Surface-Constrained Solvent Model for Macromolecules

The PDLD model described above is quite useful for evaluating electrostatic free energies in proteins. However, with more computer power one can use

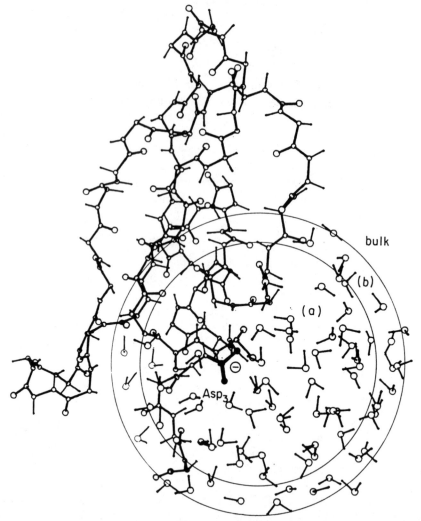

FIGURE 4.8. A surface constrained all-atom (SCAAS) model for solvated proteins. The figure depicts the different regions of the model around Asp 3 in the protein BPTI. Region *a* includes the solute atoms and the unconstrained protein atoms as well as the unconstrained water molecules. Region *b* is the surface constraints region which is surrounded by a bulk region (see Ref. 10 for more details).

more explicit solvent models. In particular, one can exploit the all-atom solvent model of Chapter 3 for electrostatic calculations of macromolecules (Ref. 10). This model, which is illustrated in Fig. 4.8, surrounds a given region in the protein by a sphere of water molecules, deleting all molecules that are within a van der Waals distance from the protein atoms. Next, the solvent molecules within the surface region are constrained to have the

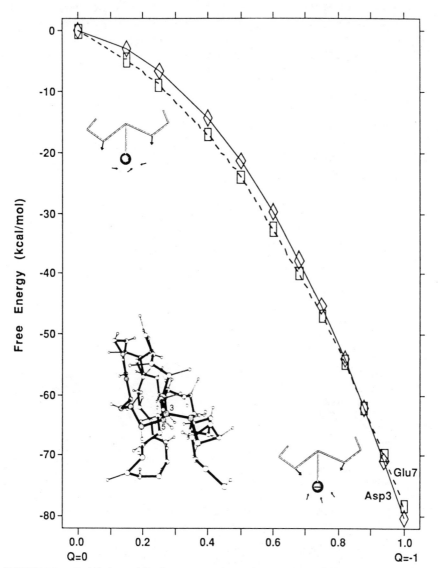

FIGURE 4.9. Caclulations of the free energy of charging Asp 3 and Glu 7 in BPTI. The figure describes the calculated "solvation" free energy [obtained with the use of eq. (3.20)] as a function of the charge of the corresponding acid. The same type of calculations for a charge in water are described in Fig. 3.2 and demonstrated by Program 3.B.

polarization and radial constraint they would have in the corresponding infinite system.

This explicit model is particularly useful for free-energy perturbation (FEP) calculations in proteins. For example, if we like to evaluate the energy associated with charging a given group in a protein (see Refs. 10 and 11) we can use eq. (3.20) and perform the same type of calculations described in Section 3.3.2, but with the actual protein microenvironment rather than pure water. Such calculations are described in Fig. 4.9 for the free energy of charging the residues Asp 3 and Glu 7 in their corresponding sites in the protein BPTI. The same type of model can be used for FEP/EVB calculations of chemical reactions in proteins, using the approach described in Section 3.5 (see Chapter 5).

4.7. SOME RELEVANT COMPUTER PROGRAMS

4.A. Steepest Descent Minimization Program

```
c       d          The first derivatives vector ( i.e. the gradient of the potetial energy )
c       e          The energy of the system
c       ib,jb      The bonded atoms
c       nat        The number of atoms
c       nb         The number of bonds
c       nsteep     The number of iterations
c       x          The coordinate vector
        real x(nsteep,3 ),d(nsteep,3 ),ens(nsteep),steps(nsteep)
        common/topo/nat,nb,ib(20),jb(20)
        read*,nat,nsteep
        do i=1,nat
            read*,(x(i,j),j=1,3)
        enddo
        read*,nb
        do i=1,nb
            read*,ib(i),jb(i)
            write(6,*) ib(i),jb(i)
        enddo
        do ith=1,nsteep
            call molec(x,d,e)
            call steepd(d,x,e,ith,ens,steps,step)
        enddo
        end
c==========================================================
        subroutine steepd(d,x,e,ith,ens,steps,step)

c       this subroutine evaluates the coordinate shift
c       by using a simple steepest descent method.
        real x(nsteep,3 ),d(nsteep,3 ),ens(nsteep),steps(nsteep)
        common/topo/n,nb,ib(20),jb(20)
        ens(ith) = e
        if(ith.eq.1) then
            step=0.005                        ! Initializes step size.
```

```
            else
                if ( ens ( ith ).gt.ens( ith - 1 ) ) then
                    step = 0.3 * step              ! If energy goes up, decrease step size.
                else
                    step = 1.2 * step              ! If energy goes down, increase step size.
                endif
            endif
            steps( ith ) = step
            s=0.0
            do i = 1 , n
                do j = 1 , 3
                    s = s + d(i,j) **2
                enddo
            enddo
            s =step/sqrt(s)
            do i = 1 , n
                do j = 1 , 3
                    x(i,j) = x(i,j) - s * d(i,j)
                enddo
                print*,'coordinate',(x(i,j),j=1,3)
            enddo
            return
            end
c===========================================================
            subroutine molec(x,d,e)
c           Calculates the energy and first derivatives
c           for the given system
c           b          The bond length
            real d(nsteep,3),x(nsteep,3),c(3)
            common/topo/nat,nb,ib(20),jb(20)
            e=0.
            do i=1,n
                do k=1,3
                    d(i,k)=0.
                enddo
            enddo
            do i=1,nb
                i1=ib(i)
                j1=jb(i)
                b=0.
                do k=1,3
                    c(k)=x(i1,k)-x(j1,k)
                    b=b+c(k)**2
                enddo
                b=sqrt(b)
                call ebond(b,eb,df)
                e=e+eb
                do k=1,3
                    d(i1,k)=d(i1,k)+df*c(k)/b
                    d(j1,k)=d(j1,k)-df*c(k)/b
                enddo
            enddo
            print 100,e
            return
100         format(2x,'energy',f10.2)
            end
```

```
c================================================================
      subroutine ebond(b,f,df)

      f=50*(b-1.5)**2
      df=100*(b-1.5)
      return
      end
c================================================================
c================================================================
```

Data for (4.A) the steepest descent method

```
3 100        nat nsteep
0.0 0.0 0.0  x1 y1 z1
1.5 0.0 0.0  x2 y2 z2
0.0 1.5 0.0  x3 y3 z3
3            nb
1 2          ib  jb
2 3
3 1
```

4.B.   A Convergent Minimization Method

```
c         The modified Newton Raphson method

c         d         The first derivatives vector
c         ib,jb     The bonded atoms
c         nat       The number of atoms
c         nmin      The number of iterations
c         x         The coordinate vector
      implicit real*8 (A-H,O-Z)
      dimension F(20,20),x(20,3),d(20,3),d2(20,3)
      dimension c(20,20),ev(20)
      integer s,t
      common/topo/nat,nb,ib(20),jb(20)
      read*,nat,nmin
      do i=1,nat
          read*,(x(i,j),j=1,3)
      enddo
      read*,nb
      do i=1,nb
          read*,ib(i),jb(i)
      enddo
      do ith = 1,nmin                       ! Calculates the second derivatives
          s=0                               ! by a numerical method.
          eps = 0.02
          do j=1,nat
              do k=1,3
                  s=s+1
                  t=0
                  do l=1,nat
                      do m=1,3
                          call molec(x,d,e)
                          x(l,m)=x(l,m) - eps
```

```
                              call molec(x,d2,e)
                              x(l,m)=x(l,m) + eps
                              t=t+1
                              F(s,t)=(d(j,k)-d2(j,k))/eps
                      enddo
                  enddo
              enddo
          enddo
          print101,e
101       format(2x,'energy',f10.6)
          n=3*nat
          write(6,*) 'F matrix'
          do i=1,n
              print 1001,(F(i,j),j=1,n)
          enddo
1001      format(9f11.3)
          call diag(F,c,ev,n)
          call newton_min(n,nat,x,d,c,ev)
      enddo
      end
c==================================================================

      subroutine newton_min(n,nat,x,d,c,ev)

c     This subroutine performs a modified Newton Raphson minimization

c     c         the eigenvectors of the second derivatives matrix
c     d         the corresponding gradient
c     x         the n dimensional coordinate vector
      implicit real*8 (A-H,O-Z)
      dimension x(20,3),d(20,3),c(20,20),ev(20),d1(20),s(20)
      l=0
      do i=1,nat
          do k=1,3
              l=l+1
              s(l)=0.
              d1(l)=d(i,k)
          enddo
      enddo
      sumd2 = 0.0
      do i = 1, n
          sumd2 = sumd2 + d1(i)**2
      enddo
      sumd2 = sqrt(sumd2/n)
      stepi = 0.1
      do i=1,n
      if(i.eq.1) then
          cr0 = 0.0004*ev(1)
      endif
      sum=0.0
      do k=1,n
          sum=sum+d1(k)*c(k,i)
      enddo
      if(dabs(ev(i)).ge.cr0) then
          write(6,*) 'i ev ',i,ev(i)
          ev(i)=1./ev(i)                              ! Filters small eigenvalues
```

```
                    shift=sum*ev(i)
                    if(ev(i).lt.0) shift = (stepi/sumd2)*sum
                    do j=1,n
                         s(j)=s(j)-shift*c(j,i)
                    enddo
              endif
         enddo
         l=0
         do i=1,nat
              do k=1,3
                   l=l+1
                   x(i,k)=x(i,k)+s(l)
              enddo
         enddo
         do i = 1,nat
              write(6,*) 'coord. - ',(x(i,j),j=1,3)
         enddo
         return
         end
c===============================================================
         subroutine molec(x,d,e)

         The corresponding subroutine is given in 4.A.
c===============================================================
         subroutine ebond(b,f,df)

         implicit real*8 (A-H,O-Z)
         f=50*(b-1.5)**2
         df=100*(b-1.5)
         return
         end
c===============================================================
         .subroutine DIAG

         The corresponding subroutine is given in 1.A.
c===============================================================
c===============================================================
```

Data for (4.B) the Newton Raphson method

```
2  3               nat,nmin
0.0 0.0 0.0        X1,Y1,Z1
1.7 0.0 0.0        X2,Y2,Z2
1                  nb
1  2               ib,jb
```

4.C.   Normal Mode Analysis

```
c       d          The first derivatives vector
c       ib,jb      The bonded atoms
c       kb         The force constant
c       nat        The number of atoms
c       nb         The number of bonds
```

```
c       x          The coordinate vector
c       y(i)       The atomic masses
        implicit real*8 (A-H,O-Z)
        dimension w(20),y(20),G(20,20)
        dimension F(20,20),x(20,3),d(20,3),d2(20,20)
        dimension c(20,20),ev(100)
        common/topo/nat,nb,ib(20),jb(20),kb(20)
        integer s,t
        s = 0
        cons = 1.d3*2.0455/(2.99793*2*3.141592)
        read*,nat
        t = 1
        do i=1,nat
            read*,(x(i,j),j=1,3)
            read(5,*) y(i)
            w(t)=y(i)
            w(t+1)=y(i)
            w(t+2)=y(i)
            t = t + 3
        enddo
        read*,nb
        do i=1,nb
            read*,ib(i),jb(i),kb(i)
        enddo
        eps=1.0d-3
        do j=1,nat                              ! Calculates the second derivative of the
            do k=1,3                            ! potential function.
                s=s+1
                t=0
                do l=1,nat
                    do m=1,3
                        x(l,m)=x(l,m) + eps
                        call molec(x,d,e)
                        x(l,m)=x(l,m) -2*eps
                        call molec(x,d2,e)
                        x(l,m)=x(l,m) + eps
                        t=t+1
                        F(s,t)=(d(j,k)-d2(j,k))/(2*eps)
                    enddo
                enddo
            enddo
        enddo
        n=3*nat
        write(6,*) 'F matrix'
        do i=1,n
            print 1001,(F(i,j),j=1,n)
        enddo
        write (6,*) 'mass scaled F'
        do i=1,n
            do j=1,n
                g(i,j)=f(i,j)/(dsqrt(w(i))*dsqrt(w(j)))
            enddo
        enddo
        do i=1,n
```

```
            print 1001,(G(i,j),j=1,n)
        enddo
        call diag(g,c,ev,n)
        write(6,*) 'Frequency'
        do i=1,n
            ev(i)=dsqrt(dabs(ev(i)))*cons        ! The frequences ( in cm⁻¹) .
            print 1000,ev(i)
        enddo
1000    format(2x,f16.4)
        write(6,*) 'Normal mode'
        do i=1,n
            print 1002,(c(j,i),j=1,n)
        enddo
1001    format(9f8.3)
1002    format(9f8.3)
        end
c==========================================================
        subroutine DIAG

        The corresponding subroutine is given in 1.A.
c==========================================================
        subroutine molec(x,d,e,kf)

        The corresponding subroutine is given in 4.A.
c==========================================================
        subroutine ebond(b,f,df,ki)

        implicit real*8 (A-H,O-Z)
        f=ki*(b-1.5d0)**2
        df=(ki*2)*(b-1.5d0)
        return
        end
c==========================================================
c==========================================================
```

Data for(4.C)normal modes of water

```
3               nat
-.75  0. 0.     x1 y1 z1
1.0             mass1
0.75  0. 0.     x2 y2 z2
1.0             mass2
0.0 1.299 0.0   x3 y3 z3
16.0            mass3
3               nb
1 2 75          ib jb fk
2 3 535
3 1 535
```

REFERENCES

1. (a) S. Lifson and A. Warshel, *J. Chem. Phys.*, **49**, 5116 (1986); (b) A. Warshel and S. Lifson, *J. Chem. Phys.*, **53**, 582 (1970); (c) A. Warshel, M. Levitt, and S. Lifson, *J. Mol. Spectrosc.*, **33**, 84 (1970); (d) A. Warshel in *Modern Theoretical Chemistry*, Vol 7, G. A. Segal (Ed.), (1977), p. 133, Plenum, New York.

2. N. L. Allinger, *Adv. Phys. Org. Chem.*, **13**, 1 (1976).

3. R. Fletcher, *Practical Methods of Optimization*, Wiley, New York, 1980.

4. A. Jack and M. Levitt, *Acta Cryst.*, **A34**, 931 (1978).

5. E. B. Wilson, J. C. Decius, and P. C. Cross, *Molecular Vibrations*, McGraw-Hill, New York, 1955.

6. D. A. McQuarrie, *Statistical Mechanics*, Harper and Row, New York, 1976.

7. M. Levitt, *J. Mol. Biol.*, **168**, 595 (1983).

8. (a) A. Warshel and S. T. Russell, *Quart. Rev. Biophys.*, **17**, 283 (1984). (b) S. T. Russell and A. Warshel, *J. Mol. Biol.*, **185**, 389 (1985). (c) A. Warshel and M. Levitt, *J. Mol. Biol.*, **103**, 227 (1976)

9. C. Tanford and J. G. Kirkwood, *J. Am. Chem. Soc.*, **79**, 5333 (1957).

10. A. Warshel, F. Sussman, and G. King, *Biochemistry*, **25**, 8368 (1986).

11. A. Warshel, *Pontif. Acad. Sci. Script. Var.*, **55**, 59 (1984).

12. C. F. Wong and J. A. McCammon, *J. Am. Chem. Soc.*, **108**, 3830 (1986).

13. S. N. Rao, U. C. Singh, P. A. Bash and P. A. Kollman, *Nature*, **328**, 551 (1987).

14. A. Warshel and J. Aquist, *Ann. Rev. Biophys. Biophys. Chem.*, **20**, 267 (1991).

15. K. A. Sharp and B. Honig, *Ann. Rev. Biophys. Biophys. Chem.*, **19**, 301 (1990).

5

MODELING REACTIONS IN ENZYMES: AN INTRODUCTION

This chapter will outline the main principles of modeling enzymatic reactions and leave to subsequent chapters the examination of specific enzymes and different catalytic factors.

The basic approach of treating reactions in the microenvironment of enzyme-active sites is identical to the one taken in modeling reactions in solutions. However, in studying enzymes we will have an advantage over the calculations of solution reactions; that is, the enzyme potential surfaces will be *calibrated* by using solution reactions as reference systems. Thus we will view the enzyme as another "solvent" and focus on the *changes* in the EVB potential surfaces moving from the reference solvent (i.e., water) to the enzyme-active site. In the actual simulations and discussions we will examine two approaches: (1) The simplified PDLD method, which evaluates free energies using the average protein coordinates as obtained from X-ray crystallography, while modeling the solvent by the LD model, and (2) an explicit all-atom model for both the protein and the solvent which evaluates free energies by exploring the phase space of the system.

5.1. ENZYME KINETICS AND FREE-ENERGY SURFACES

5.1.1. Basic Concepts of Enzyme Kinetics

Since a significant part of our discussion will involve comparison of reactions in solutions and in proteins it is important to establish a link between experimental kinetic measurements in such systems and the corresponding free-energy surfaces.

Typical single-substrate enzymatic reactions can be described by the kinetic scheme (see Refs. 1 and 2 for more extensive discussions).

$$E + S \underset{k_{-1}}{\overset{k_1}{\rightleftharpoons}} ES \overset{k_{cat}}{\rightarrow} EP \rightarrow E + P \tag{5.1}$$

where S and P are the substrate and product, respectively. Such reactions are usually monitored in terms of the substrate concentrations $[S]$ and are analyzed in terms of the velocity.

$$v = -d[S]/dt \tag{5.2}$$

The use of the steady-state approximation gives this velocity as a function of $[S]$ by

$$v = ([S]/([S] + K_M))[E_T]k_{cat} \tag{5.3}$$

where E_T is the total enzyme concentration and the constant K_M is given by

$$K_M = (k_{-1} + k_{cat})/k_1 \tag{5.4}$$

When k_{cat} is smaller than k_{-1} we can approximate K_M by the equilibrium constant K_S for the dissociation of the substrate from the enzyme

$$K_M \geq K_S = [E][S]/[ES] = k_{-1}/k_1 \tag{5.5}$$

The dependence of v on $[S]$ follows saturation kinetics, as shown in Fig. 5.1. The asymptotic value v_{max} provides a convenient estimate of k_{cat} by

$$v_{max} = k_{cat}[E_T] \tag{5.6}$$

At low substrate concentrations the velocity is given by (Ref. 1)

$$v = (k_{cat}/K_M)[E_T][S] \tag{5.7}$$

Thus we can consider (k_{cat}/K_M) as an apparent second-order rate constant. This constant is the most critical parameter in determining the specificity of

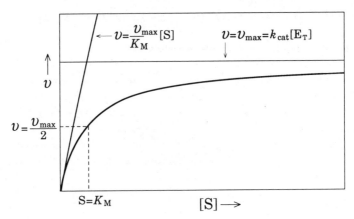

FIGURE 5.1.   The relationship between v and $[S]$ in a typical enzymatic reaction.

the enzyme for its substrate. Because the enzyme and the substrate cannot combine more rapidly than diffusion permits, there is an upper limit on enzyme catalysis and k_{cat}/K_M cannot be larger than $10^9 \, s^{-1} \, M^{-1}$.

5.1.2. The Relationship Between Enzyme Kinetics and Apparent Activation Free Energy

The meaning of k_{cat}/K_M can best be understood by considering the free-energy diagram of Fig. 5.2. As seen from the figure, the activation barrier for the reaction, $\Delta g^{\neq}$, is given by

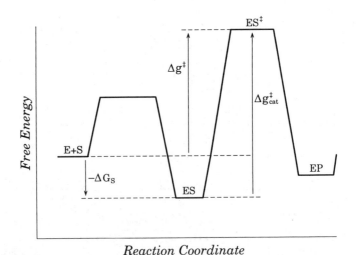

Reaction Coordinate

FIGURE 5.2.   A schematic free-energy diagram for a typical enzymatic reaction.

$$\Delta g^{\neq} = \Delta g_{cat}^{\neq} - \Delta G_s \qquad (5.8)$$

When the condition $K_M \simeq K_S$ is satisfied, we have [from eq. (5.8), eq. (3.31), and the relationship $K_s = \exp\{-\Delta G_s \beta\}$]

$$\ln(k_{cat}/K_M) = \ln(\bar{\tau}^{-1}) - (\Delta g_{cat}^{\neq} - \Delta G_s)\beta$$
$$= \ln(\bar{\tau}^{-1}) - \Delta g^{\neq}\beta \qquad (5.9)$$

Thus the apparent rate constant (k_{cat}/K_M) is determined by the apparent activation barrier $\Delta g^{\neq}$. In fact, both $\Delta g^{\neq}$ and $\Delta g_{cat}^{\neq}$ should have been written as $\Delta G^{\neq}$ and $\Delta G_{cat}^{\neq}$, respectively [see eq. (2.11)], but as long as we do not have large entropic effects (see Chapter 9), the approximation given above is reasonable.

For further discussion and for calibration purposes we will have to define a reference solution reaction which would be sufficiently related to the given enzymatic reaction. This will help us in removing the problems associated with concentration effect and in focusing on the actual catalytic advantage of the enzyme. To do this we describe the reference reaction in solution as

$$(S) + (R) \underset{k_{-1}}{\overset{k_1}{\rightleftharpoons}} (SR)_{cage} \overset{k_{cage}^{\neq}}{\rightarrow} (PR')_{cage} \rightarrow (R') + (P) \qquad (5.10)$$

where S, P and R designate, respectively, the substrate, the product and the reactive groups that participate in the actual chemical reaction in the enzyme. Here we separate the solution reaction to the stage involved in bringing the reacting fragments to the same solvent cage and the subsequent reaction. The free-energy profile associated with this treatment is drawn in

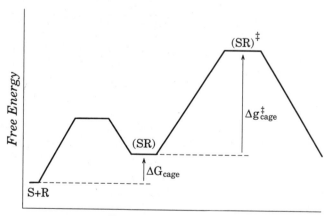

FIGURE 5.3.   A schematic free-energy profile for a reference reaction in a solvent cage.

Fig. 5.3. Now the reaction rate is determined by ΔG_{cage} and $\Delta g_{cage}^{\neq}$, but ΔG_{cage} is almost entirely determined by simple concentration factors. Thus a comparison of $\Delta g_{cage}^{\neq}$ and $\Delta g_{cat}^{\neq}$ allows one to explore fundamental catalytic aspects, including real entropic effects, without preoccupation with the rather trivial "effective concentration" effect, associated with bringing the reactants to the same cage.

Exercise 5.1. Evaluate ΔG_{cage}^{0} for the reaction $A + B \rightarrow AB$ in water. Assume that the size of both A and B is the same as that of a water molecule and that the cage volume is equal to that of six water molecules.

Solution 5.1. As will be explained in Chapter 9, the free energy contribution associated with confining a molecule that occupies a volume v_1 to a volume v_2 is given by $\Delta G^{0} = -\beta^{-1} \ln(v_2/v_1)$. Since we deal with a standard condition (i.e., one molar of A and B) we start our problem with one molecule of A and one molecule of B in a volume $v_1 = (1/N)_{liter} \approx 1600 \text{ Å}^3$, where N is the Avogadro's number. Since the molarity of water is 55 we can also take v_1 as the volume of 55 water molecules ($v_1 = 55v_w$). In evaluating ΔG_{cage}^{0} we can fix B and ask what is the free energy of confining A to a volume $v_2 = v_{cage} = 6v_w$ around B. This gives $\Delta G_{cage}^{0} = -\beta^{-1} \ln(v_2/v_1) = -\beta^{-1} \ln(6v_w/55v_w) \approx 1.3 \text{ kcal/mol}$.

5.2. PDLD SIMULATIONS OF PROTON TRANSFER REACTIONS IN ENZYMES

5.2.1. The His–Cys Ion Pair in Papain as a Model System

The class of proton transfer (PT) reactions plays a major role in many biological processes, including various enzymatic reactions. This class of reactions will be served here as a general example and an introduction for more complicated reactions. As a specific demonstration let's consider a proton transfer between Cys 25 and His 159 in papain. This reaction can be formally described as

$$(Im + R - SH)_p \rightarrow (ImH^+ + R - S^-)_p \tag{5.11}$$

where Im designates histidine (or imidazole) residue.

The simplest way to consider the energetics of this reaction is to use the EVB model for the reacting region and PDLD model for the protein. The EVB potential surface is formally identical to that used in Chapter 2, where the reaction is described in terms of two resonance structures,

$$\psi_1 = Im \quad H - S$$

$$\psi_2 = Im^+ - H \quad S^- \tag{5.12}$$

where we use the notation ψ rather than Φ [(see eq. (2.40)] since we implicitly mix the high energy $\text{Im H}^+ \text{S}^-$ state with our two states. The potential surface of the reactive region (which will be referred to as the "solute") is described by

$$\varepsilon_1^0 = H_{11}^0 = \Delta M(b_1) + U_{nb}^{(1)} + U_{QQ}^{(1)} + U_{strain}^{(1)}$$

$$\varepsilon_2^0 = H_{22}^0 = \Delta M(b_3) + U_{nb}^{(2)} + U_{QQ}^{(2)} + \alpha_2^0 + U_{strain}^{(2)}$$

$$H_{12}^0 = A \exp\{-\mu(r_2 - r_2^0)\} \tag{5.13}$$

where b_1, b_3, and r_2 are respectively the S–H, N–H and S$\cdots$N distances; ΔM is a Morse potential (relative to its minimum) and U_{nb} is the nonbonded interaction within the atoms of the given resonance structure. $U_{QQ}^{(i)}$ is the electrostatic interaction between the fragments of the ith resonance form and is given (in kcal/mol) by

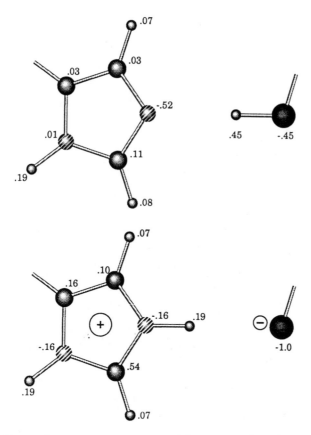

FIGURE 5.4. The two key resonance structures and the corresponding atomic charges for PT in papain.

TABLE 5.1. EVB Parameters for PT from X–H to Im[a,b]

Bonds	$\Delta M = D[1 - \exp\{-a(b - b_0)\}]^2$		
O–H	$D = 102$	$b_0 = 0.96$	$a = 2.35$
S–H	$D = 81$	$b_0 = 1.32$	$a = 2.35$
N^+–H	$D = 93$	$b_0 = 1.00$	$a = 2.35$
C–O	$D = 93$	$b_0 = 1.43$	$a = 2.06$
C–S	$D = 70$	$b_0 = 1.80$	$a = 2.00$

Bond Angles	$U_\theta = \frac{1}{2} K_\theta (\theta - \theta_0)^2$	
N–H–X	$K_\theta = 5$	$\theta_0 = 3.141$
C–X–H	$K_\theta = 100$	$\theta_0 = 1.911$

Nonbonded[c]	$U_{nb} = A \exp\{-ar\}$	
$N \cdots H$	$A = 150$	$a = 2.5$
$O^- \cdots H$	$A = 65$	$a = 2.5$
$S^- \cdots H$	$A = 300$	$a = 2.5$

Nonbonded	$U'_{nb} = A_i A_j r_{ij}^{-12} - A_i A_j r_j^{-6}$	
C	$A = 632$	$B = 20$
O	$A = 632$	$B = 24$
O^-	$A = 1400$	$B = 24$
S	$A = 632$	$B = 24$
S^-	$A = 3000$	$B = 24$
N	$A = 774$	$B = 24$
H	$A = 4$	$B = 0$

Charges	$U_{qq} = 332 q_i q_j / r_{ij}$	
$(S - H)(\psi_1)$	$q_S = -0.45$	$q_H = +0.45$
$(O - H)(\psi'_1)$	$q_O = -0.45$	$q_H = +0.45$
$(S^-)(\psi_2)$	$q_S = -1.0$	
$(O^-)(\psi'_2)$	$q_O = -1.0$	
$(Im)(\psi_1)$	see Fig. 5.4	
$(ImH^+)(\psi_2)$	see Fig. 5.4	

Off diagonal parameters and diagonal shifts		
H_{12}	$A_{12}^{NS} = -140$ $r^0 = 0.0$	$\mu_{12}^{NS} = 0.6$
α_1	0.0	
α_2	148	

[a] Energies in kcal/mol, distance in Å, angles in radians and charges in au. Parameters not listed can be taken from Table 4.2.

[b] X designates S or O atoms.

[c] The potential U_{nb} is used for nonbonded interactions between the indicated EVB atoms, while U'_{nb} is used for nonbonded interactions between all other atoms, as long as the corresponding atoms are not bonded to each other or to a common atom.

$$U_{QQ}^{(i)} = 332 \sum_{m>m'} \sum_{\mu>\mu'} Q_{m\mu}^{(i)} Q_{m'\mu'}^{(i)} / r_{m\mu,m'\mu'} \tag{5.14}$$

where m runs over fragments, μ over atoms, the Q's are the atomic charge in au, and the r's are the indicated distances in Å. U_{strain} describes the bond-stretching potential for the bonds which are not being broken during the reaction as well as the angle-bending and torsion potentials for the solute system. The contributions to U_{strain} are represented by standard force-field expressions and are given by

$$U_{\text{strain}}^{(i)} = \frac{1}{2} \sum_m K_{b,m}^{(i)} (b_m^{(i)} - b_{0,m}^{(i)})^2 + \frac{1}{2} \sum_m K_{\theta,m}^{(i)} (\theta_m^{(i)} - \theta_{0,m}^{(i)})^2$$
$$+ \sum_m K_{\phi,m}^{(i)} \cos(n^{(i)}\phi_m) + \sum_m K_{\chi,m}^{(i)} (\chi_m^{(i)} - \chi_{0,m}^{(i)})^2 \tag{5.15}$$

where the angles and torsion parameters depend on the hybridization of the atoms in the given fragments. The χ term represents the potential for out-of-plane deformation of sp^2 atoms (e.g., the histidine carbons). The notation for the other terms is the same as before. The charges $Q^{(i)}$ are given in Fig. 5.4 and the relevant parameters for the various terms in eq. (5.15) are given in Table 5.1.

5.2.2. Calibrating the Enzyme Surface Using Solution Experiments

The energetics of the solute state in the protein site can be expressed as

$$\varepsilon_i^p = \varepsilon_i^0 + \Delta g_{\text{sol},p}^{(i)}$$

$$H_{ij}^p = H_{ij}^0 \tag{5.16}$$

where p designates protein and $\Delta g_{\text{sol},p}^{(i)}$ is the solvation energy of the charges $Q^{(i)}$ of the ith resonance structure by the protein + water system. The only difference between eq. (5.16) and the corresponding equation [eq. (2.31)] for the same reaction in water is that $\Delta g_{\text{sol},p}^{(i)}$ replaces $\Delta g_{\text{sol},w}^{(i)}$. This offers a unique chance for calibration using the experimental information about the reaction in water. That is, for proton transfer reactions in water we need to evaluate the parameter α_2^0 from the not so reliable gas-phase calculations or from the frequently unavailable gas-phase experiments. For the reaction in protein, on the other hand, we can rewrite eq. (5.16) as (Ref. 3)

$$\varepsilon_i^p = \varepsilon_i^w + (\Delta g_{\text{sol},p}^{(i)} - \Delta g_{\text{sol},w}^{(i)}) = \varepsilon_i^w + (\Delta\Delta g_{\text{sol}}^{(i)})_{w\to p} \tag{5.17}$$

ε_i^w can be easily calibrated by using eq. (2.34) and writing

$$\alpha_2^0 \simeq (\Delta G_{PT,w}^\infty)_{\text{obs}} - \Delta\Delta g_{\text{sol},w}^{2,\infty} \tag{5.18}$$

where we use here the notation $\Delta\Delta g_{sol,w}^{i,\infty}$ to designate the corresponding $(\Delta g_{sol,w}^{i,\infty} - \Delta g_{sol,w}^{1,\infty})$ and $(\Delta G_{PT,w}^{\infty})_{obs}$ is obtained from the relationship.

$$(\Delta G_{PT,w}^{\infty})_{obs} = 1.38(pK_a^w(SH) - pK_a^w(Im\,H^+)) \simeq 6\,kcal/mol$$

(5.19)

This gives a first approximation for α_2^0, without any gas-phase information. The resulting α_2^0 can be used to calculate $\Delta G_{PT,w}$, which is then refined by adjusting α_2^0 until the calculated and observed values of $\Delta G_{PT,w}$ coincide. With the calibrated α_2^0 we can now evaluate the protein potential surface in terms of the difference in solvation free energy, $(\Delta\Delta g_{sol}^{(2)})_{w\to p}$, associated with the transfer of the ionic configuration from a solvent cage to the protein-active site. To see this point, let us start with the PDLD potential surface for the reaction in solution. A simplified version of this surface is shown in Fig. 5.5. The figure presents the so-called *potential of mean force* for the reaction, without addressing the probability of moving the proton donor and acceptor to the same solvent cage. This is, however, sufficient for

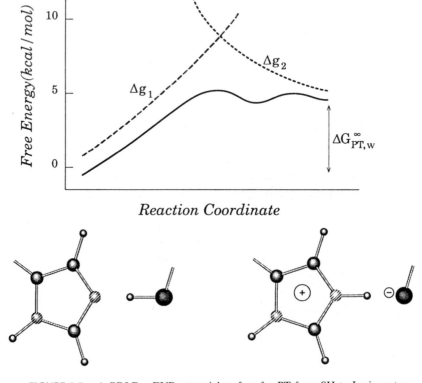

FIGURE 5.5. A PDLD + EVB potential surface for PT from SH to Im in water.

the purpose of comparing the reaction in the reference solvent cage to the corresponding reaction in the active site.

Next we evaluate the PDLD + EVB surface for the enzymatic reaction using eq. (5.17). The resulting surface is shown in Fig. 5.6. As seen from the figure, the protein can reduce $\Delta g^{\neq}$ by stabilizing the ionic state more than water. In fact, in the specific case of papain the protein inverts the stabilization of the covalent and ionic states relative to their order in solution.

An interesting point that emerges from Fig. 5.6 is the relation between $\Delta g^{\neq}$ and $(\Delta \Delta g_{sol})_{w \to p}$. As seen from the figure, the lowering of the activation energy for the reaction is almost linearly proportional to the stabilization of the ionic resonance form $(\Delta \Delta g_{sol}^{(2)})_{w \to p}$. An enzyme which is designed to accelerate a proton transfer between A and B will simply stabilize the $(B^{+}-H \quad A^{-})$ state more than water.

Exercise 5.2. (a) Use the noniterative PDLD of Program 2.B to evaluate the energetics of a proton transfer reaction $Im + CH_3SH \to Im\,H^{+} + CH_3S^{-}$ where the N and S atoms are 3 Å apart in water. To simplify the calculations use only three atoms (N, H and S) as the reacting atoms. Use the calculations to determine the parameter α_2^0 that reproduces the observed proton transfer energetics in water. This parameter will, of course, be different than the one in Table 5.1 since the Im residue is replaced by a single nitrogen atom. (b) With the adjusted α_2^0 calculate the proton transfer energy between the same two groups in a naive model of a "protein" composed of a single, positively charged group 3 Å from the S^{-} ion. (c) Repeat the same calculations for papain, assigning all the histidine charges to the proton-accepting nitrogen and including all the protein dipoles whose

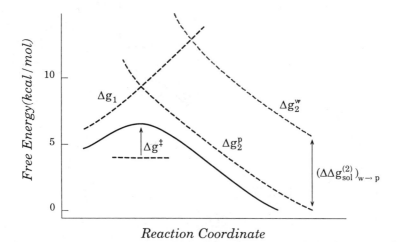

FIGURE 5.6. A PDLD + EVB surface for PT from Cys-25 to Im-159 in papain.

closest distance to the reacting groups is less than 3 Å (define these dipoles by assigning standard residual charges to the protein atoms). Surround this "truncated" protein system by an LD solvent model.

Solution 5.2. (a) Use Program 5.A. (c) Use Program 5.B.

5.3. ALL-ATOM MODELS FOR PROTON TRANSFER REACTIONS IN ENZYMES

Although the physical picture of the previous section is quite appealing, one can argue that use of eq. (5.17) is based on an ad hoc assumption and that a more rigorous treatment is needed. This can be readily provided by the use of an all-atom model for the protein–solvent system and the free-energy perturbation approach. That is, we now represent the two diagonal energies by potential functions rather than free energies, using

$$\varepsilon_i^p = \varepsilon_i^0 + U_{Qq,Sp}^{(i)} + U_{nb,Sp}^{(i)} + U_{ind,Sp}^{(i)} + U_{pp}$$

$$H_{ij}^p = H_{ij}^0 \qquad (5.20)$$

where the notation is similar to that of eq. (3.12) but now the solvent, s is replaced by the protein p. For example, consider a proton transfer from Ser 195 to His 57 in the active site of trypsin. The solute EVB potential surfaces ε_i^0 are identical to those in eq. (5.13), but with the serine OH replacing the cysteine SH. The potential surfaces ε_i^p are then obtained by eq. (5.20) and mixed to give the ground state surface (Refs. 3 and 4):

$$E_g^p = \frac{1}{2} \{(\varepsilon_1^p + \varepsilon_2^p) - [(\varepsilon_1^p - \varepsilon_2^p)^2 + 4H_{12}^2]^{1/2}\} \qquad (5.21)$$

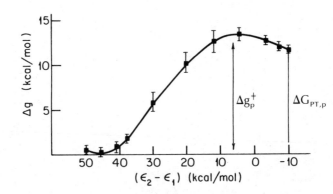

FIGURE 5.7. A FEP surface for PT between serine and histidine in trypsin (the calculations are taken from Ref. 4).

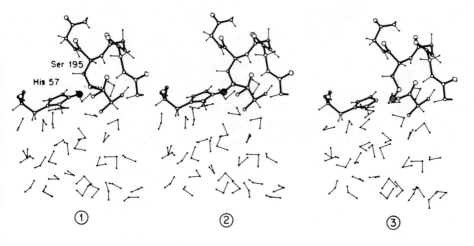

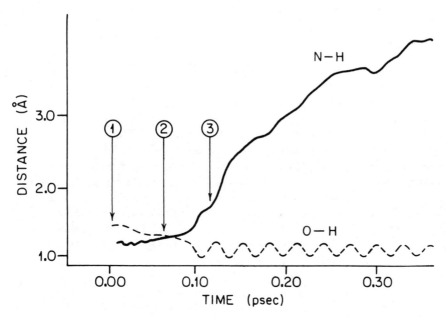

FIGURE 5.8. A downhill trajectory for the proton transfer step in the catalytic reaction of trypsin. The trajectory moves on the actual ground state potential, from the top of the barrier to the relaxed enzyme–substrate complex. 1, 2, and 3 designate different points along the trajectory, whose respective configurations are depicted in the upper part of the figure. The time reversal of this trajectory corresponds to a very rare fluctuation that leads to a proton transfer from Ser 195 to His 57.

This ground state provides an analytical potential surface for the reacting system in the enzyme-active site.

Exercise 5.3. Construct an EVB potential surface for a proton transfer from Cys-25 to an active site of water molecules in papain.

The analytical surfaces ε_1^p and ε_2^p and the corresponding analytical ground state surface E_g can be conveniently used in FEP studies of chemical reactions in solutions. Such calculations can be performed in a complete analogy to the calculations presented in Section 3.5, using eq. (3.29) to evaluate the free energy surface for the proton transfer reaction. The corresponding results were evaluated in ref. 4 and are summarized in Fig. 5.7. The calculations appear to give converging results with a rather small error range. This and subsequent studies (e.g., Ref. 3c) have indicated that such calculations can be used in comparing the activation energy of the reaction in the enzyme-active site ($\Delta g^{\neq}$) to the corresponding reaction in a solvent cage ($\Delta g_{cage}^{\neq}$).

The analytical potential surface E_g can also be used for examination of dynamical effects. This can be done in the same way as outlined in Chapter 3, propagating trajectories downhill from the transition state and calculating $\bar{\tau}$ of eq. (3.31). A typical downhill trajectory for our reaction is shown in Fig. 5.8.

5.4. LINEAR FREE-ENERGY RELATIONSHIP IN ENZYMES

The ability to obtain converging results from all-atom free-energy calculations should allow one to examine the linear free-energy relationship assumed in the simplified PDLD + EVB approach. That is, in using eq. (5.17) it was assumed that the free energies of the VB states can be substituted in the EVB Hamiltonian and give an approximated free-energy surface for the ground state of the reacting system. Obviously, this approximation is not exact nor can it be proven by any first principle derivation. However, if computer experiments show that $\Delta g^{\neq}$ is linearly related to ΔG, we can adopt the simplified approximation of eq. (5.17). Test calculations of proton-transfer reactions in proteins can be easily accomplished by a parametric change of α_2^0 and evaluation of the corresponding $\Delta g_{PT}^{\neq}$ and $\Delta G_{PT,p}$ (here we use the notation $\Delta G_{PT,p}$ in analogy with the corresponding notation for proton transfer in solution). Such calculations, which are presented in Fig. 5.9, strongly support the validity of linear free-energy relationships in proteins (Refs. 3b and 4a). Experimental verifications of this point are also emerging from recent works (Refs. 5 and 6). This explains why the simplified model, that only considers the effect of the environment on the diagonal elements of eq. (5.17), gives similar results to those obtained by all-atom FEP calculations.

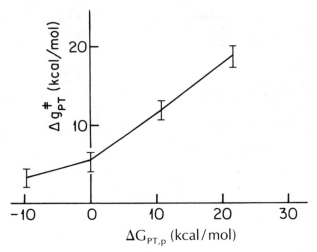

FIGURE 5.9. The relationship between $\Delta g^{\neq}$ and ΔG for proton transfer reactions in the active site of trypsin.

The main lesson from the analysis given above is that the activation free energy of the reaction is strongly correlated with the stabilization of the ionic resonance structure by the protein-active site. The generality of this concept will be considered in the following chapters.

5.5. SOME RELEVANT COMPUTER PROGRAMS

5.A. A Simplified EVB Calculation of a Proton Transfer between Cysteine and Histidine in Water. (The reaction is represented in an oversimplified way by assigning all the histidine residual charges to a single nitrogen)

Use program 2.B

Data for 5.A

```
'C' 'Cl'
92. 69. 1.87 1.43 5200. 2.5 100          ! The potential parameters.
'H' 'F'
134. 61. 2.27 0.92 640. 2.5 20.
'H' 'Cl'
102. 78. 1.90 1.28 1000. 2.5 25.
'H' 'Br'
89. 70. 1.82 1.41 2000. 2.5 30.
'H' 'S'
81. 73. 1.87 1.32 100. 2.5 975.
'H' 'O'
102. 75. 2.26 0.96 100. 2.5 975.
'H' 'I'
```

```
74. 61. 1.75 1.60 1950. 2.5 6400.
'H' 'C'
106. 96. 1.80 1.09 2000. 2.5 30.
'H' 'N'
103. 78. 2.07 1.00 10. 2.5 50.
'C' 'N'
89. 70. 1.45 1.41 2000. 2.5 30.
'C' 'S'
70. 60. 1.87 1.7  2000. 2.5 30.
'*' '*'
3   3          natom, nr
1
2  3
1
1  2
0
'N' 'H' 'S'
'N' 'H' 'S'
'N' 'H' 'S'
0.000  0.200 -0.200                              ! The solute charges
0.800  0.200 -1.000
0.000  1.000 -1.000
0.000  135.  265
28.376 69.648 23.355                             ! The solute coordinates
27.716 70.291 22.743
26.297 70.833 21.486
0
0.000
```

5.B. A Simplified EVB Calculation of Proton Transfer between Cys 25 and His 159 in Papain. (The solute is represented in an oversimplified way by assigning all the histidine residual charges to a single nitrogen)

use program 2.B

Data for 5.B

```
'C' 'Cl'
92. 69. 1.87 1.43 5200. 2.5 100.
'H' 'F'
134. 61. 2.27 0.92 640. 2.5 20.
'H' 'Cl'
102. 78. 1.90 1.28 1000. 2.5 25.
'H' 'Br'
89. 70. 1.82 1.41 2000. 2.5 30.
'H' 'S'
81. 73. 1.87 1.32 100. 2.5 975.
'H' 'O'
102. 75. 2.26 0.96 100. 2.5 975.
'H' 'I'
74. 61. 1.75 1.60 1950. 2.5 6400.
'H' 'C'
```

```
106. 96. 1.80 1.09 2000. 2.5 30.
'H' 'N'
103. 78. 2.07 1.00 10. 2.5 50.
'C' 'N'
89. 70. 1.45 1.41 2000. 2.5 30.
'C' 'S'
70. 60. 1.87 1.7 2000. 2.5 30.
'*' '*'
3    3
3    3          natom, nr
1
2    3
1
1    2
0
'N' 'H' 'S'
'N' 'H' 'S'
'N' 'H' 'S'
0.000   0.200  -0.200
0.800   0.200  -1.000
0.000   1.000  -1.000
0.000   135.  265
28.376 69.648  23.355
27.716 70.291  22.743
26.297 70.833  21.486
 76
31.121 72.151  22.889  0.500
30.905 71.153  22.262  -0.500
30.082 72.683  23.828  -0.834
30.287 73.468  24.429  0.417
29.199 72.235  23.922  0.417
27.166 73.774  25.252  -0.194
26.334 73.281  25.508  0.097
27.938 73.149  25.187  0.097
26.959 74.531  24.000  0.550
25.863 74.608  23.543  -0.550
28.092 75.238  23.352  -0.400
29.002 75.239  23.752  0.400
27.850 75.981  22.111  -0.050
27.020 76.534  22.186  0.050
29.051 76.825  21.998  0.050
28.808 77.722  21.631  0.050
29.725 76.384  21.410  0.050
29.464 76.927  23.129  -0.650
30.208 77.624  23.153  0.500
27.718 74.966  21.026  0.550
26.951 75.230  20.151  -0.550
28.310 73.606  21.286  0.000
29.058 73.533  21.943  0.000
27.849 72.358  20.599  0.000
28.549 71.650  20.724  0.000
```

```
27.655 72.604  19.113  0.550
28.599 72.840  18.401  -0.550
26.568 72.013  21.373  0.000
25.795 72.475  20.949  0.000
26.670 72.381  22.295  0.000
26.251 72.540  18.599  -0.400
25.549 72.325  19.288  0.400
25.791 72.713  17.189  -0.097
25.974 71.872  16.683  0.097
24.302 72.960  17.211  -0.194
23.827 72.127  17.495  0.097
24.022 73.193  16.280  0.097
23.918 74.102  18.125  0.000
23.928 74.119  19.632  -0.097
24.085 73.331  20.225  0.097
23.662 75.506  20.114  -0.450
23.665 75.825  21.079  0.450
28.115 71.109  14.102  0.050
28.052 70.216  13.659  0.050
27.298 71.657  13.923  0.050
28.316 70.992  15.266  -0.650
29.096 70.329  15.355  0.500
22.243 73.334  22.846  -0.194
21.956 72.495  23.305  0.097
23.178 73.261  22.497  0.097
27.534 66.767  21.487  -0.194
27.367 65.881  21.926  0.097
28.306 66.680  20.887  0.097
26.684 67.591  19.353  0.550
27.034 66.675  18.669  -0.550
```

```
27.866 67.710  22.545  0.000
27.071 67.850  23.786  0.000
26.368 67.237  24.159  0.000
27.688 69.011  24.411  0.000
27.665 69.345  25.345  0.000
28.987 68.688  22.530  0.000
29.278 69.125  21.679  0.000
26.605 69.013  18.851  -0.400
26.456 69.775  19.511  0.400
27.007 69.345  17.473  -0.097
26.859 70.312  17.307  0.097
28.454 69.074  17.584  0.550
29.067 69.591  18.482  -0.550
36.501 68.240  24.585  -0.194
37.172 68.374  25.315  0.097
36.672 67.348  24.168  0.097
35.127 68.248  25.191  0.000
33.855 68.830  24.635  -0.097
33.763 69.361  23.794  0.097
32.734 68.474  25.566  -0.450
31.762 68.747  25.462  0.450
```

The coordinates and
residual charges of the
protein atoms around
the His -Cys pair in Papain.

REFERENCES

1. A. Fersht, *Enzyme Structure and Mechanism*, W. H. Freeman, Reading and San Francisco, 1977.
2. W. P. Jencks, *Catalysis in Chemistry and Enzymology*, Dover Publications, New York, 1987.
3. (a) A. Warshel, *Biochemistry*, **20**, 3167 (1981); (b) A. Warshel, *Pontif. Acad. Sci. Script. Var.*, **55**, 59 (1984); (c) A Warshel, F. Sussman, and J-K. Hwang, *J. Mol. Biol.*, **201**, 139 (1988).
4. A. Warshel, S. Russell, and F. Sussman; Israel, *J. Chem.*, **27**, 217 (1986).
5. A. R. Fersht, R. J. Leatherbarrow, and T. N. C. Wells, *Nature*, **322**, 284 (1986).
6. M. D. Toney and J. F. Kirsch, *Science*, **243**, 1485 (1989).

GENERAL ACID CATALYSIS AND ELECTROSTATIC STABILIZATION IN THE CATALYTIC REACTION OF LYSOZYME

6.1. BACKGROUND

Hen egg-white lysozyme catalyzes the hydrolysis of various oligosaccharides, especially those of bacterial cell walls. The elucidation of the X-ray structure of this enzyme by David Phillips and co-workers (Ref. 1) provided the first glimpse of the structure of an enzyme-active site. The determination of the structure of this enzyme with trisaccharide competitive inhibitors and biochemical studies led to a detailed model for lysozyme and its hexa N-acetyl glucoseamine (hexa-NAG) substrate (Fig. 6.1). These studies identified the C–O bond between the D and E residues of the substrate as the bond which is being specifically cleaved by the enzyme and located the residues Glu 37 and Asp 52 as the major catalytic residues. The initial structural studies led to various proposals of how catalysis might take place. Here we consider these proposals and show how to examine their validity by computer modeling approaches.

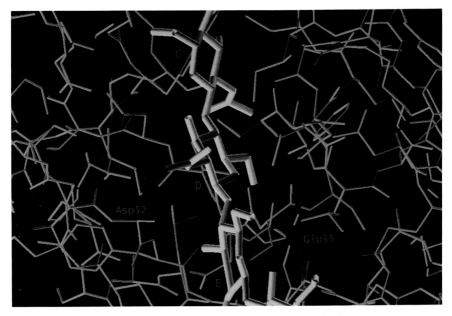

FIGURE 6.1. The lysozyme molecule and its bound substrate.

In analyzing various proposed catalytic effects we will have to specify an assumed mechanism and examine its energetics. Comparing the relative contributions to $(\Delta g_p^{\neq} - \Delta g_{cage}^{\neq})$ from different catalytic factors should tell us which of them are really important. Of course, if the assumed mechanism is incorrect we will be asking a somewhat irrelevant question.

The commonly accepted mechanism for the catalytic reaction of lysozyme is the so-called *general acid catalysis* mechanism.

$$\text{ROR}' + \text{AH} \rightleftharpoons \text{ROH}^+\text{R}' + \text{A}^- \rightleftharpoons \text{R}^+ + \text{A}^- + \text{R}'\text{OH} \overset{\text{H}_2\text{O}}{\rightleftharpoons} \text{ROH} + \text{AH} + \text{R}'\text{OH}$$
$$(6.1)$$

where in our case R and R' are sugar residues. The assumed rate-limiting step of this reaction (shown in Fig. 6.2) consists of a proton transfer from Glu-35 to O_4 and a cleavage of the protonated $C-O_4$ bond, forming a transition state with a positively charged carbon atom which is referred to as a *"carbonium ion transition state."* To examine the ways the protein catalyzes this reaction, let us consider several feasible proposals.

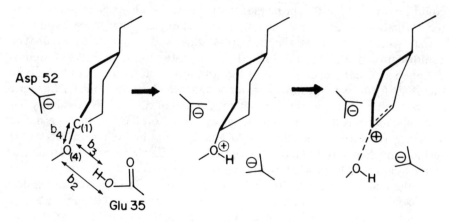

FIGURE 6.2.   A schematic description of the rate-determining steps in the catalytic reaction of lysozyme.

6.2.   THE STRAIN HYPOTHESIS AND PROTEIN FLEXIBILITY

In order to change the rate constant of a given reaction, it is crucial that the enzyme will recognize some *change* in the reacting system. Such a change

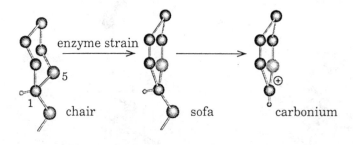

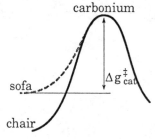

FIGURE 6.3.   The strain hypothesis for the catalytic reaction of lysozyme. The enzyme is assumed to destabilize the chair geometry by pushing the ground state substrate toward the sofa configuration. This ground-state destabilization effect is supposed to reduce $\Delta g_{cat}^{\neq}$.

could involve, at least in principle, the geometrical changes associated with the formation of the carbonium transition state; the sugar ring assumes a planar sofa-like form at the transition state and a nonplanar chair-type conformation at the ground state. If, as suggested in Fig. 6.3, the protein is designed as a steric compliment of the sofa form and it does not like to bind the chair form, we will have a ground state which is destabilized by steric strain. Such a situation will result in a reduction of $\Delta g_{cat}^{\neq}$. This possibility has been advanced as a major catalytic factor, based on the fact that the native structure of the enzyme appears to compliment the sofa form of the D residue much better than that of the chair form. However, the structural studies cannot tell us what is the energy associated with the deformation from the chair to the sofa forms in the D subsite. Here the recommended approach is not to ask whether the strain is a possible catalytic factor, but to find out its actual contribution to $\Delta g_{cat}^{\neq}$. To do this we have to model the relevant aspects of the reaction. This will be done in the next section by the full EVB formulation, but in addressing the strain problem we may restrict

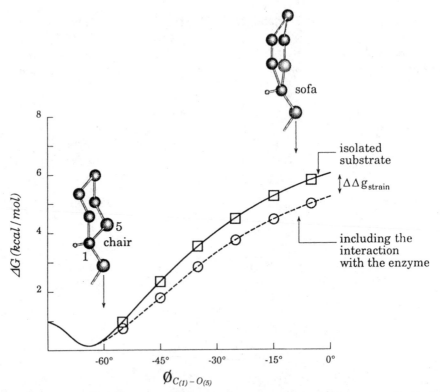

FIGURE 6.4. Examining the strain hypothesis by a free-energy perturbation study. The results are plotted as a function of ϕ rather than λ, for simplicity. The ΔG is given relative to the reference potential $[-K\cos(\phi_{C_1-O_5})]$. The figure demonstrates that the strain effect is not significant.

ourselves to a much simpler treatment, comparing the free energy associated with the chair→sofa transformation in water to that in the protein site. This can be done by a free-energy perturbation method (see Chapter 3), using a mapping potential of the form

$$\varepsilon_m = \varepsilon_1(1 - \lambda_m) + [-K\cos(\phi_{C_1-O_5}) + \varepsilon_1\lambda_m \qquad (6.2)$$

where ε_1 is the ground state potential of the substrate and its surrounding, K is a large constant and the $-K\cos\phi$ term has a minimum at $\phi = 0$. Changing the mapping parameter λ_m from zero to one forces the sugar to change its structure from the chair to the sofa configuration and gives the corresponding free-energy change by eq. (3.19). Figure 6.4 summarizes such a calculation. As seen from the figure, it costs some free energy, $\Delta\Delta g^{\neq}_{\text{strain}}$, for the enzyme active site to deform and accommodate the change from the chair to the sofa conformation. However, this strain contribution is *small* (about 1 kcal/mol). The reason for this small effect is intuitively obvious. That is, as illustrated in Fig. 6.5, the overall atomic displacements associated

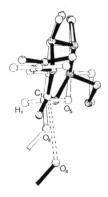

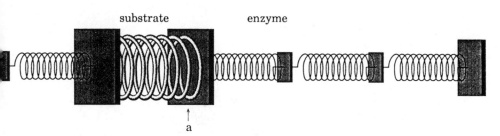

FIGURE 6.5. Illustrating why strain energy is not a key catalytic factor. As argued in the text, the small Cartesian displacements of the substrate in the chair→sofa deformation (upper part of the figure) can be distributed over the many degrees of freedom of the protein. This point is best understood by considering the substrate as a spring with large force constant while considering the protein as an array of coupled springs with small force constants.

with the chair→ sofa transition are quite small, if one superimposes the two structures in a way that minimizes the shift in Cartesian coordinates and the corresponding response of the protein. The protein, with its many bond-stretching and angle-bending degrees of freedom, can easily accommodate small Cartesian shifts without storing a large amount of strain energy. This point can be considered intuitively by describing the protein as a collection of springs (lower part of Fig. 6.5) that can undergo a significant displacement for a small cost in energy, by distributing a small part of the displacement over each spring. The same type of conclusions are obtained from simpler energy minimization studies (Ref. 2). In fact, it one could build a mechanical model of balls and springs for the enzyme substrate complex, he would have seen that the flexible enzyme cannot deform the substrate, nor store a large tension upon substrate displacements.

Exercise 6.1. To illustrate the small cost associated with a total deformation of $0.5\,Å$ by a collection of bonds, evaluate the energy involved in compressing point a of Fig. 6.5 by $0.5\,Å$ to the left while distributing the resulting strain in the three springs, whose energy can be described by $U_i = \frac{1}{2} K \Delta b_i^2$ with $K = 30\,\text{kcal/mol}^{-1}\,Å^{-2}$.

Solution 6.1. The least-energy accommodation of the $0.5\,Å$ shift will be obtained by distributing it equally over the three springs. This gives $\Delta U = 3 \times (30/2) \times (0.166)^2 \simeq 1.2\,\text{kcal/mol}$. A smaller value would be obtained with more springs.

In view of the considerations given above it appears that strain energy cannot be a major catalytic factor as long as we deal with regular reactions where the geometrical changes associated with the formation of the transition state do not exceed $1\,Å$.

6.3. MODELING CHEMISTRY AND ELECTROSTATIC EFFECTS

6.3.1. A Simple VB Formulation

Inspection of the active site of lysozyme suggests the possibility that electrostatic effects might be important. That is, the negatively charged Asp-52 group is situated in a position where it can stabilize the positively charged carbonium transition state (Ref. 3). However, experiments with model compounds in solutions (Ref. 4), which are depicted schematically in Fig. 6.6, show no major catalytic effect due to a properly situated negative charge. This reason led many to discard electrostatic effects as a major catalytic factor. However, the strength of electrostatic interaction in the interior of proteins may be drastically different than the corresponding strength in solution since the local microscopic dielectric effect could be very different. An oversimplified macroscopic attempt to estimate the dielectric

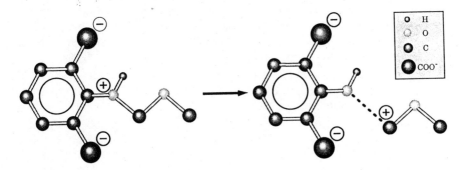

FIGURE 6.6. The type of model compounds that were used to estimate the electrostatic stabilization in lysozyme (the only hydrogen atom shown, is the one bonded to the oxygen). Such molecules *do not* show a large rate acceleration due to electrostatic stabilization of the positively charged carbonium transition state. However, the reaction occurs in solution and not in a protein-active site, and the dielectric effect is expected to be very different in the two cases.

constant inside the protein-active site (see exercise 6.2) from the observed effect of Asp 52 on the pK_a Glu 35 indicated that the effect of Asp 52 on the transition state is small (Ref. 5).

Exercise 6.2. Chemical substitution experiments have indicated that the presence of the negatively charged Asp 52 changes the pK_a of Glu 35 by 1.1 units. Using the distances between Asp 52 and Glu 35 and between Asp 52 and C_1 (which are 6.2 and 3.8 Å, respectively) and a uniform dielectric constant, estimate the stabilization of C_1^+ by Asp 52.

Solution 6.2. Using Coulomb's law for both the Asp $\cdots$ Glu interaction and the Asp$^- \cdots C^+$, interaction, we have $\Delta G_{Asp\cdots Glu} = 332/(r \times \varepsilon) = 332/(6.2 \times \varepsilon) = 1.38 \Delta pK_a = 1.52$, which gives $\varepsilon \approx 35$. Using this ε for the Asp $\cdots C_1$ interaction we obtain $\Delta G_{Asp^-C_1^+} \simeq -332/(3.8 \times 35) = 2.5$ kcal/mol. This is a significant effect, but far too small to account for the observed rate enhancement by the enzyme, which leads to more than 7 kcal/mol change in the activation free energy.

One may suggest that the enzyme has a smaller dielectric effect than the one deduced from the above exercise and that this leads to a large electrostatic effect. Unfortunately, Asp52 would not be ionized in an active site with a low dielectric constant (charged groups are not stable in low dielectric environments as demonstrated in Ref. 8a of Chapter 4). Thus, we may conclude, in agreement with the above exercise, that the dielectric constant for charge-charge interaction in the active site of lysozyme is large and that electrostatic stabilization is *not* a major catalytic effect. However, the above arguments are based on oversimplified macroscopic considerations and, as was pointed out in Chapter 4, the dielectric effect in proteins cannot be assessed by Coulomb's law type macroscopic models. This is particularly true when one deals with the fundamental problem of the magnitude of the electrostatic contribution to catalysis. The electrostatic problem is far too important to be left as a macroscopic exercise with an assumed dielectric constant and must be addressed by explicit microscopic molecular models, such as those developed in Section 4.6.

In order to really assess the magnitude of the electrostatic effect in lysozyme on a microscopic level it is important to simulate the actual assumed chemical process. This can be done by describing the general acid catalysis reaction in terms of the following resonance structures:

$$\psi_1 = (\text{A–H} \quad \text{R—O—R}')$$

$$\psi_2 = (\text{A}^- \quad \overset{\overset{\displaystyle H}{|}}{\text{R—O}^+\text{—R}'})$$

$$\psi_3 = (\text{A}^- \quad \text{R}^+ \overset{\overset{\displaystyle H}{|}}{\text{O—R}'})$$

$$\psi_4 = (\text{A–H} \quad \text{R}^+ \text{O—R}') \tag{6.3}$$

In order to construct free-energy surfaces for this system we start by defining the diagonal matrix elements, or the "force fields", for each resonance structure:

$$\varepsilon_1 = \Delta M(b_1) + \Delta M(b_4) + U_{nb}^{(1)} + U_{QQ}^{(1)} + U_{strain}^{(1)} + U_{Ss}^{(1)} + U_{ss}$$

$$\varepsilon_2 = \Delta M(b_4) + \Delta M(b_3) + U_{nb}^{(2)} + U_{QQ}^{(2)} + U_{strain}^{(2)} + \alpha_2^0 + U_{Ss}^{(2)} + U_{ss}$$

$$\varepsilon_3 = \Delta M(b_3) + U_{nb}^{(3)} + U_{strain}^{(3)} + U_{QQ}^{(3)} + \alpha_3^0 + U_{Ss}^{(3)} + U_{ss}$$

$$\varepsilon_4 = \Delta M(b_1) + U_{nb}^{(4)} + U_{strain}^{(4)} + U_{QQ}^{(4)} + \alpha_4^0 + U_{Ss}^{(4)} + U_{ss}$$

$$H_{ij} = H_{ij}^0 = A_{ij} \exp\{-\mu(r_{kl(ij)} - r_{ij}^0)\} \tag{6.4}$$

where b_1, b_2, b_3, and b_4 are, respectively, the A–H, A–O, H–O, and R–O bond lengths (see Fig. 6.7). ΔM is a Morse-type function for the indicated bond (relative to its minimum value), U_{nb} is the nonbonded interaction for the given resonance structure and U_{strain} is given by

$$U_{strain}^{(i)} = \frac{1}{2} \sum_m K_{b,m}^{(i)}(b_m^{(i)} - b_{0,m}^{(i)})^2 + \frac{1}{2} \sum_m K_{\theta,m}^{(i)}(\theta_m^{(i)} - \theta_{0,m}^{(i)})^2$$

$$+ \sum_m K_{\phi,m}^{(i)} \cos(n^{(i)}\phi_m^{(i)}) \tag{6.5}$$

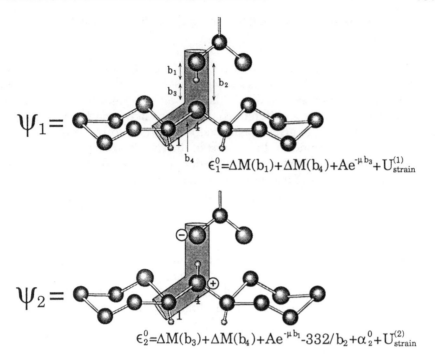

$$\Psi_1 = \qquad \epsilon_1^0 = \Delta M(b_1) + \Delta M(b_4) + Ae^{-\mu b_3} + U_{strain}^{(1)}$$

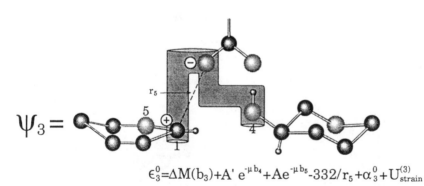

$$\Psi_2 = \qquad \epsilon_2^0 = \Delta M(b_3) + \Delta M(b_4) + Ae^{-\mu b_1} - 332/b_2 + \alpha_2^0 + U_{strain}^{(2)}$$

$$\Psi_3 = \qquad \epsilon_3^0 = \Delta M(b_3) + A'\,e^{-\mu b_4} + Ae^{-\mu b_5} - 332/r_5 + \alpha_3^0 + U_{strain}^{(3)}$$

FIGURE 6.7. The key resonance structures for the catalytic reaction of lysozyme. The ϵ_i's include only the solute contributions and the complete expression is given in eqs. (6.4) and (6.5). The quantum mechanical atoms are enclosed within the shaded region.

where the angles and torsion parameters depend on the given hybridization of the central carbon atom in the R group (e.g., sp^2 for ψ_3 and ψ_4). This strain force field keeps the equilibrium structure of the R^+ fragment in the sofa configuration and that of R in ψ_1 and ψ_2 in the chair configuration (see Fig. 6.7). The terms $U_{QQ}^{(i)}$, U_{Ss}, and U_{ss} are defined in Chapter 5 and the atom pair kl used for the off-diagonal element are chosen according to the specific H_{ij}.

6.3.2. Calibrating the EVB Surface Using the Reference Reaction in Solution

In order to make an effective use of the VB formulation we have to calibrate the relevant parameters using reliable experimental information. The most important task is to obtain the relevant α_i^0. Since the α's represent the energy of forming the different configurations in the gas phase at infinite separation between the given fragments, it is natural to try to obtain them from gas-phase experiments. In the case of the catalytic reaction of lysozyme one can compile the relevant information from the available gas-phase experiments (Table 6.1) and use it to determine the α's.

For example, we can estimate α_3^0 by

$$\alpha_3^0 \simeq \varepsilon_3^0(\infty) - \varepsilon_1^0(\infty) \simeq \Delta H_{\text{gas}}^{3,\infty} - \Delta H_{\text{gas}}^{1,\infty} \tag{6.6}$$

where the ε^0 do not include any solvent contribution. Using this expression, we obtain $\alpha_3^0 \sim 167$ kcal/mol. However, in many cases it is not simple to find gas-phase experiments about charged fragments and, as indicated in Chapter 5, it is frequently more convenient to obtain the α's from solution experi-

TABLE 6.1. Gas-Phase Enthalpies that Can Be Used to Determine the Energies of the Different Configurations Involved in the Catalytic Reaction of Lysozyme[a]

Entry	Process	Expression Used	ΔH kcal/mol
1	$R'OR \rightarrow R'O^- + R^+$	$D + I - EA$	215
2	$R'OH \rightarrow R'O^- + H^+$	$D + I - EA$	376
3	$HCOOH \rightarrow HCOO^- + H^+$	$D + I - EA$	345
4	$HCOOH + H_2O \rightarrow HCOO^- + H_3O^+$	ΔH_{PTg}	177
5	$CH_3OH + H_2O \rightarrow CH_3O^- + H_3O^+$	ΔH_{PTg}	211
6	$HCOO^- + R'OH \rightarrow HCOOH + R'O^-$	ΔH_{PTg}	44
7	$HCOOH + ROR' \rightarrow HCOO^- + ROH^+R'$	ΔH_{PTg}	147
8	$HCOOH + ROR' \rightarrow HCOO^- + R^+ + R'OH$	$\Delta H_{PTg} + \Delta H$	167
9	$ROH^+R' \rightarrow R^+ + R'OH$	ΔH	20
10	$ROH^+R' \rightarrow R + R'OH^+$	$\Delta H - I_R + I_{ROH}$	76
11	$ROH^+R' \rightarrow RO^+R' + H$	$PA_{ROR'} - I_H + I_{ROR'}$	97

[a]Information compiled in Ref. 6, where R and R' are typical C_3H_7 and C_2H_5 groups. See Ref. 6 for more details about the different notations.

ments than from gas-phase studies. That is, one can use eq. (2.34) and write

$$\alpha_i^0 \approx (\Delta G_{i,w}^\infty) - \Delta\Delta g_{sol,w}^{i,\infty} \tag{6.7}$$

where $\Delta\Delta g_{sol,w}^{i,\infty}$ is the indicated solvation energy (in water) relative to the solvation energy of state 1. This can be conveniently used for the determination of α_2^0 for the proton transfer configuration. The corresponding procedure is identical to the one used in Chapter 5 and is given here as an example.

Exercise 6.3.   Estimate α_2^0 from the relevant pK_a values and the approximate solvation energies. Use $(pK_a(Glu) = 4;\ pK_a(R-OH^+-R') = -5$

$$\Delta G_{sol,w}(GluH \rightarrow Glu^-) \simeq -70\,kcal/mol$$

$$\Delta G_{sol,w}(ROR' \rightarrow RO'(H^+)R') \simeq -65\,kcal/mol$$

Solution 6.3. Using eqs. (2.32), (2.34) and (6.7) we obtain $\Delta G_{2,w}^\infty \approx 12\,kcal/mol;\ \alpha_2^0 = 147\,kcal/mol$.

In considering ε_3 we note that α_3^0 is already known (eq. 6.6) and we may use it to obtain $\Delta G_{3,w}^\infty$ rather than the other way around. With $\alpha_3^0 = 167\,kcal/mol$ and the solvation energies of the various fragments, one obtains (Ref. 6) $\Delta G_{3,w}^\infty = 26 \pm 5\,kcal/mol$. We also obtain readily

$$\Delta G_{4,w}^\infty = {}^*\!\Delta G_{3,w}^\infty + 2.3\,RT(pK_a(R'OH) - pK_a(AH)) \approx 41 \pm 5 \tag{6.8}$$

and $\alpha_4^0 = 215\,kcal/mol$.

Finally, we should also exploit one more key experimental fact—the activation barrier for the dissociation of the R–O bond in the protonated R–OH$^+$R' molecule is available from kinetic studies of the so-called "specific acid catalysis" reaction.

$$ROR' + H_3O^+ \overset{k_1}{\rightleftharpoons} ROH^+R' + H_2O \overset{k_2}{\rightleftharpoons} R^+ + R'OH + H_2O \tag{6.9}$$

where the acid is an hydronium ion. An analysis of these studies gives (Ref. 6) $k_2 \sim 1 - 10s^{-1}$, which yields through eqs. (3.31) and (2.12) an activation barrier of about 18 kcal/mol. Thus we can use the estimate $(\Delta g_{2\rightarrow3}^{\neq})_w^\infty = 18\,kcal/mol$, where the superscript ∞ indicates that A$^-$ is at infinite separation from the protonated C–O bond. These experimental estimates are summarized in Table 6.2.

With these ΔG^∞ we can estimate the energetics of the key asymptotic point on the potential surface of the reference reaction in which AH and R–O–R' are kept in the same solvent cage. First, we note that (ΔG_2) is

TABLE 6.2. Experimental Determination of the Energies (in kcal/mol) at the Asymptotic Points of the Potential Surface of the General Acid Catalysis Reaction[a,b]

Configuration	Notation	Expression Used	ΔG_w	ΔH_{gas}
$A^- + RO^+HR'$	$\Delta G^\infty_{2,w}$	$2.3\,RT[pK_a(AH)$ $-pK_a(RO^+HR')]$	12 ± 2	147 ± 5
$A^- + R^+ + R'OH$	$\Delta G^\infty_{3,w}$	$\Delta G_{2,w} + (\Delta G_{2\to3})_w$	26 ± 5	167 ± 5
$AH + R^+ + R'O^-$	$\Delta G^\infty_{4,w}$	$\Delta G^\infty_{3,w} + 2.3\,RT[pK_a(R'OH)$ $-pK_a(AH)]$	41 ± 2	215 ± 5
$A^- + R^+ \cdots OHR'$	$(\Delta g^{\neq}_{1\to3})_w$	$\Delta G^\infty_{2,w} + (\Delta g^{\neq}_{2\to3})_w$	29 ± 2	

[a]The gas-phase ΔH values are based on analysis of gas-phase experiments, which are given in Ref. 6.
[b]See discussion in text for the evaluation of the ΔG's.

reduced by about 2 kcal/mol, when A^- and $R\text{–}OH^+\text{–}R'$ are brought together, due to the electrostatic interaction between these fragments. The activation barrier for the proton transfer step can be estimated by noting that the reverse reaction $(2 \to 1)$ is an exothermic reaction and that such proton transfer reactions are usually diffusion-controlled reactions with 5 kcal/mol or less activation barriers. Thus $(\Delta g^{\neq}_{2\to1})_w < 5$ and $(\Delta g^{\neq}_{1\to2})_w \le \Delta G_{2,w} + 5$. The barrier $(\Delta g^{\neq}_{2\to3})_w$ is expected to be similar to $(\Delta g^{\neq}_{2\to3})^\infty_w$, giving

$$(\Delta g^{\neq}_{2\to3})_w \le (\Delta g^{\neq}_{2\to3})^\infty_w - 2 \qquad (6.10)$$

The inequality indicates that *if* a concerted mechanism (where b_4 and b_2 change simultaneously) gives a $\Delta g^{\neq}$ which is much lower than our stepwise estimate, we will have smaller $\Delta g^{\neq}_{cage}$. This possibility, however, is not supported by detailed calculations (Ref. 6). Direct information about $\Delta g^{\neq}_{cage}$ can be obtained from studies of model compounds where the general acid is covalently linked to the R–O–R' molecules. However, the analysis of such experiments is complicated due to the competing catalysis by H_3O^+ and steric constraints in the model compound. Thus, it is recommended to use the rough estimate of Fig. 6.8. If a better estimate is needed, one should simulate the reaction in different model compounds and adjust the α parameters until the observed rates are reproduced.

With the estimates of Fig. 6.8 we are ready to determine the off-diagonal elements. These elements can be obtained by fitting our four-states gas-phase potential surface to the more rigorous six-states EVB surface given in ref. 6 (or to other gas-phase quantum mechanical surfaces) using the expression given in eq. (6.4).

Alternatively, one can obtain the H_{ij} by forcing the calculated solution surface to reproduce the observed information about the solution reaction. The same procedure should also be used for fine tuning the α's parameter.

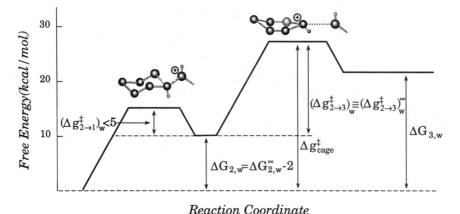

Reaction Coordinate

FIGURE 6.8. The energetics for the reference reaction in solution (see text for discussion and further clarification of the difference between our reference reaction and the actual mechanism in solution).

The various approximated H_{ij} are given in Table 6.3 together with the parameters for the diagonal matrix elements.

It should be noted at this stage that the reference reaction of Fig. 6.8 does not necessarily correspond to the actual mechanism in solution. That is, our reference reaction represents a mathematical trick that guarantees the correct calibration for the asymptotic energies of the enzymatic reaction (by using the relevant solution experiments). This may be viewed as a

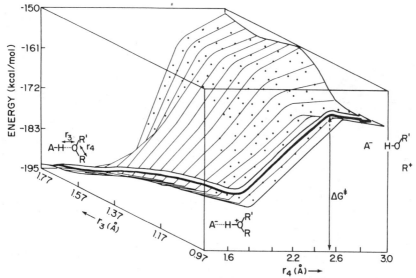

FIGURE 6.9. Potential surface for a general acid-catalysis reaction in solution; r_3 and r_4 are the O_B–H and C–O distances, respectively. Regions of the potential surface with more than 50% ionic character are dotted (see Ref. 6 for more details).

TABLE 6.3. Parameters for the Reaction of Lysozyme[a,b]

Bonds		$\Delta M = D[1 - \exp\{-a(b - b_0)\}]^2$	
O–H	$D = 102$	$b_0 = 0.96$	$a = 2.35$
C–O	$D = 92$	$b_0 = 1.43$	$a = 2.06$
C–O$^+$	$D = 76$	$b_0 = 1.43$	$a = 2.06$
O$^+$–H	$D = 97$	$b_0 = 0.96$	$a = 2.35$

Bond angles		$U_\theta = \frac{1}{2}K_\theta(\theta - \theta_0)^2$	
X–C–Y	$K_\theta = 60$		$\theta_0 = 1.911$
X–C$^+$–Y	$K_\theta = 60$		$\theta_0 = 2.094$
X–O–Y	$K_\theta = 60$		$\theta_0 = 1.911$
X–O$^+$–Y	$K_\theta = 60$		$\theta_0 = 2.094$

Nonbonded		$U_{\text{nb}} = A \exp\{-ar\}$	
O$^-\cdots$H	$A = 65$		$a = 2.5$
O$\cdots$C$^+$	$A = 5288$		$a = 2.5$
O$\cdots$H	$A = 65$		$a = 2.5$

Nonbonded		$U'_{\text{nb}} = A_i A_j r_{ij}^{-12} - B_i B_j r_{ij}^{-6}$	
C	$A = 632$		$B = 20$
C$^+$	$A = 632$		$B = 20$
O	$A = 632$		$B = 24$
O$^-$	$A = 1400$		$B = 24$
H	$A = 4$		$B = 0$

Charges		$U_{qq} = 332 q_i q_j / r_{ij}$		
$(O_a - C - O_b - H)(\psi_1)$	$q_{O_a} = -0.4$	$q_C = 0.4$	$q_{O_b} = -0.4$	$q_H = 0.4$
$(O_a - C - O_b)^-(\psi_2, \psi_3)$	$q_{O_a} = -0.7$	$q_C = 0.4$	$q_{O_b} = -0.7$	
$(O_c - C - O_d)(\psi_1)$	$q_{O_c} = -0.2$	$q_C = 0.2$	$q_{O_d} = 0.0$	
$(O_c - C - O_d^+)(\psi_2)$	$q_{O_c} = -0.2$	$q_C = 0.2$	$q_{O_d} = 0.8$	$q_H = 0.2$
$(O_c - C)^+(H - O_d)(\psi_3)$	$q_{O_c} = 0.2$	$q_C = 0.8$	$q_H = 0.5$	$q_{O_d} = -0.5$

Off diagonal parameters and diagonal shifts			
H$_{12}$	$A_{12}^{OO} = 60$	$\mu = 0.4$	$r^0 = 0.0$
H$_{13}$	$A_{13} = 0.0$	$\mu = 0.0$	$r^0 = 0.0$
A$_{23}$	$A_{23}^{CO} = 55$	$\mu = 0.5$	$r^0 = 0.0$
α_1	0.0		
α_2	147		
α_3	167		
α_4	215		

[a]Energies in kcal/mol, distance in Å, angle in radians and charges in au. Parameters not listed can be taken from Table 4.2.

[b]The function U_{nb} is used for interaction between the indicated EVB atoms while U'_{nb} is used for nonbonded interactions between other atoms which are not bonded to each other or to a common atom. The interactions between the EVB oxygens are modeled by the corresponding 6–12 potential.

practical way for obtaining the gas phase α's and H_{ij}'s while avoiding elaborated studies of entropic effects in the actual solution reaction (see Chapter 9).

A calibrated EVB + LD surface for our system in solution is presented in Fig. 6.9. With the calibrated EVB surface for the reaction in solution we are finally ready to explore the enzyme-active site.

6.3.3. Examination of the Catalytic Reaction in the Enzyme-Active Site

After the somewhat tedious parametrization procedure presented above you are basically an "expert" in the basic chemistry of the reaction and the questions about the enzyme effect are formally straightforward. Now we only want to know how the enzyme *changes* the energetics of the solution EVB surface. Within the PDLD approximation we only need to evaluate the change in electrostatic energy associated with moving the different resonance structures from water to the protein-active site.

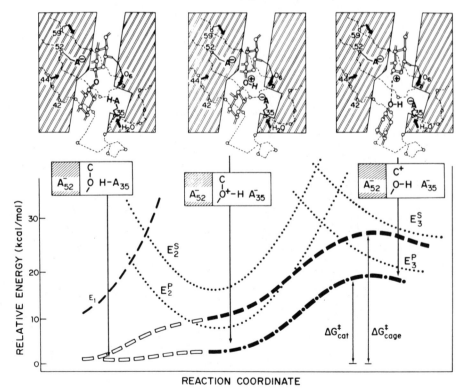

FIGURE 6.10. Comparing the energetics of the EVB configurations in solution and in the active site of lysozyme. The calculations were done by using the PDLD and related models (Refs. 6 and 7) and they represent a study of a stepwise mechanism. The energetics of a more concerted pathway (e.g., that of Fig. 6.9) is almost identical to that of the stepwise mechanism and correlated in a similar way with the electrostatic effect of the protein.

Exercise 6.4. (a) Calculate the energy of the carbonium ion configuration ψ_3 in the LD solvent model. (b) Repeat the calculations using a simplified model of the active site composed of a negative charge (that represents Asp 52) 3 Å from the C^+ atom and two *fixed* dipoles pointing toward the negative charge, in the way indicated in Fig. 6.11, while all this system is emersed in an LD solvent model.

The actual calculations that compare the energetics of the EVB configurations in the protein-active site and solutions are summarized in Fig. 6.10.

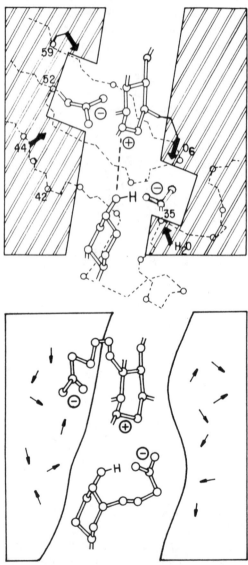

FIGURE 6.11. Comparison of the environment around the transition state of lysozyme in the enzyme-active site and in the reference solvent cage.

The calculations described in this figure produce in a qualitative way the difference in activation free energies between the reaction in the enzyme-active site and the reference solvent cage (see Refs. 6 and 7 for more details). The main reason for this catalytic effect appears to be associated with the stabilization of the positively charged carbonium ion by the negatively charged Asp 52. The effect of this group is much larger than what might be deduced from the macroscopic considerations of Exercise 6.2. Apparently the magnitude of the electrostatic stabilization effect is hard to assess without simulating the actual microscopic environment. To see this point it is instructive to view the electrostatic energetics in an alternative form, including the ionized Asp 52 in the reacting system. This is done in Fig. 6.11 which compares the transition state in the enzyme-active site to the transition state of the corresponding model compound in water. As seen from the figure, we now represent the transition state as a $(- + -)$ arrangement (e.g., Asp 52^- C^+ Glu 35^-, in the enzyme site). The enzyme manages to stabilize this system by hydrogen bonds (dipoles) which are specially aligned towards the two negatively charged acids. This gives a larger stabilization than that provided by the water dipoles to the corresponding arrangement in the reference solvent cage. The basic reason for this effect will be considered in Chapter 9.

Finally, it is important to comment that the enzyme reaction is clearly accelerated by the general acid catalysis mechanisms, since the protonation of the substrate by an acid is much more effective than that by a water molecule. This effect, however, is included in our reference reaction (e.g., the lower part of Fig. 6.11). That is, the evaluation of the concentration effect associated with bringing a glutamic acid to the same cage as the substrate is rather trivial (see Exercise 5.1) and is not the main issue in studies of enzymatic reactions. Similarly the difference between a reaction where the proton donor is an acid and a reaction where the donor is a water molecule is well understood and fully correlated with the corresponding pK_a's. The real problem is the difference between the reaction in the enzyme and in the reference solvent cage that includes all the reacting fragments, and it is here where electrostatic effects appear to be of major importance.

REFERENCES

1. (a) D. C. Phillips, *Sci. Amer.*, **215** (5), 78 (1966). (b) C. C. F. Blake, L. N. Johnson, G. A. Mair, A. C. T. North, D. C. Phillips, and V. R. Sarma, *Proc. Roy. Soc. Ser. B.*, **167**, 378 (1967).

2. A. Warshel and M. Levitt, *J. Mol. Biol.*, **103**, 227 (1976).

3. C. A. Vernon, *Proc. Roy. Soc. Ser. B*, **167**, 389 (1967).

4. B. M. Dunn and T. C. Bruice, *Adv. Enzymol. Relat. Areas Mol. Biol.*, **37**, 1 (1973).

5. J. A. Thoma, *J. Theor. Biol.*, **44**, 305 (1974).

6. A. Warshel and R. M. Weiss, *J. Am. Chem. Soc.*, **102**, 6218 (1980).

7. A. Warshel, *Biochemistry*, **20**, 3167 (1981).

7

SERINE PROTEASES AND THE EXAMINATION OF DIFFERENT MECHANISTIC OPTIONS

7.1. BACKGROUND

The serine proteases are the most extensively studied class of enzymes. These enzymes are characterized by the presence of a unique serine amino acid. Two major evolutionary families are presented in this class. The bacterial protease *subtilisin* and the *trypsin* family, which includes the enzymes *trypsin*, *chymotrypsin*, *elastase* as well as *thrombin*, *plasmin*, and others involved in a diverse range of cellular functions including digestion, blood clotting, hormone production, and complement activation. The trypsin family catalyzes the reaction:

$$- NH - \overset{\overset{\displaystyle R}{|}}{CH} - \overset{\overset{\displaystyle O}{\|}}{C} - NH - \overset{\overset{\displaystyle R'}{|}}{CH} - + H_2O \rightarrow$$

$$-NH - \overset{\overset{\displaystyle R}{|}}{CH} - CO_2H + H_2N - \overset{\overset{\displaystyle R'}{|}}{CH} - \qquad (7.1)$$

The actual reaction mechanism is very similar for the different members of the family, but the *specificity* toward the different side chain, R, differs most strikingly. For example, trypsin cleaves bonds only after positively charged Lys or Arg residues, while chymotrypsin cleaves bonds after large hydrophobic residues. The specificity of serine proteases is usually designated by labeling the residues relative to the peptide bond that is being cleaved, using the notation

$$H_2O + P_4 - P_3 - P_2 - P_1 - P_1' - P_2' \rightarrow$$

$$P_4 - P_3 - P_2 - P_1 - OH + H - P_1' - P_2' - \qquad (7.2)$$

The sensitivity of the relevant rate constants to the groups at the different sites is demonstrated in Table 7.1. The cleavage of amides in the active site of serine protease can be described formally by the two successive steps:

$$
\underset{R-C-X}{\overset{O}{\|}} + E - OH \underset{k_{-1}}{\overset{k_1}{\rightleftharpoons}} H^+ + R - \underset{X}{\overset{O^-}{\underset{|}{\overset{|}{C}}}} - O - E \underset{k_{-2}}{\overset{k_2}{\rightleftharpoons}} R - \underset{}{\overset{O}{\|}} C - O - E + HX
$$

$$
R - \overset{O}{\overset{\|}{C}} - O - E + HY \underset{k_{-3}}{\overset{k_3}{\rightleftharpoons}} R - \underset{Y}{\overset{O^-}{\underset{|}{\overset{|}{C}}}} - O - E + H^+ \underset{k_{-4}}{\overset{k_4}{\rightleftharpoons}} R - \overset{O}{\overset{\|}{C}} - Y + E - OH
$$

$$(7.3)$$

The first step, which is called the *acylation reaction*, involves a formation of an *acyl-enzyme* where the $RC(O^-)X$ group is covalently bound to the specially active serine residue and the XH group is expelled from the active site. The second step, which is called the *deacylation step*, involves an attack of an HY group on the acyl-enzyme. Here we concentrate on the acylation step which is the reverse of the second step when X and Y are identical.

The elucidation of the X-ray structure of chymotrypsin (Ref. 1) and in a later stage of subtilisin (Ref. 2) revealed an active site with three crucial groups (Fig. 7.1)–the active serine, a neighboring histidine, and a buried aspartic acid. These three residues are frequently called the *catalytic triad*, and are designated here as Asp_c His_c Ser_c (where c indicates a catalytic residue). The identification of the location of the active-site groups and intense biochemical studies led to several mechanistic proposals for the action of serine proteases (see, for example, Refs. 1 and 2). However, it appears that without some way of translating the structural information to reaction-potential surfaces it is hard to discriminate between different alternative mechanisms. Thus it is instructive to use the procedure introduced in previous chapters and to examine the feasibility of different

TABLE 7.1.   Kinetic Parameters for the Hydrolysis of Different Peptides by Elastase and Chymotrypsin

Substrate	Elastase[a]		
	k_{cat} (s^{-1})	K_M (mM)	k_{cat}/K_M $(s^{-1} M^{-1})$
Ac–Pro–Ala–NH$_2$	0.007	100	0.07
Ac–Ala–Pro–Ala–NH$_2$	0.09	4.2	21
Ac–Pro–Ala–Pro–Ala–NH$_2$	8.5	3.9	2200
Ac–Ala–Pro–Ala–Pro–Ala–NH$_2$	5.3	3.9	1360
Ac–Pro–Ala–Pro–Gly–NH$_2$	0.1	22	5
Ac–Pro–Ala–Pro–Val–NH$_2$	6.0	35	208
Ac–Pro–Ala–Pro–Leu–NH$_2$	3.0	11	270
Ac–Pro–Ala–Pro–Ala–Gly–NH$_2$	26	4.0	6500
Ac–Pro–Ala–Pro–Ala–Ala–NH$_2$	37	1.5	24700
Ac–Pro–Ala–Pro–Ala–Phe–NH$_2$	18	0.64	28800
Ac–Pro–Ala–Pro–Ala–Ala–Ala–NH$_2$	—	—	—

Substrate	Chymotrypsin[b]		
	k_{cat} (s^{-1})	K_M (mM)	k_{cat}/K_M $(s^{-1} M^{-1})$
Ac–Tyr–NH$_2$	0.17	32	5
Ac–Tyr–Gly–NH$_2$	0.64	23	28
Ac–Tyr–Ala–NH$_2$	7.5	17	440
Ac–Pro–Tyr–Gly–NH$_2$	0.76	15	51
Ac–Phe–NH$_2$	0.06	31	2
Ac–Phe–Gly–NH$_2$	0.14	15	10
Ac–Phe–Ala–NH$_2$	2.8	25	114
Ac–Pro–Phe–Gly–NH$_2$	0.76	15	51

[a]From R. C. Thompson and E. R. Blout, *Biochemistry*, **12**, 51 (1973) and C. A. Bauer et al. *Eur. J. Biochem.*, **120**, 289 (1981).
[b]From W. K. Baumann, S. A. Bizzozero, and H. Dutler, *Eur. J. Biochem.*, **39**, 381 (1973).

mechanisms. We will concentrate here on the two most likely mechanisms, which are described in Fig. 7.2.

Mechanism *a* involves a proton transfer from Ser$_c$ to His$_c$ and a nucleophilic attack of the ionized Ser$_c$ on the carbonyl carbon of the substrate, forming a negatively charged intermediate which is referred to as the *tetrahedral intermediate* (to indicate the sp^3 tetrahydral geometry around the carbon) or the *oxyanion intermediate*. Here we will designate the tetrahydral intermediate by the notation t^-. In the next stage the protonated His$_c$ donates its proton to the amide nitrogen and facilitates the departure of the H$_2$N–CHR$'$– group, leading to the formation of the acyl-enzyme. In related reactions of amide hydrolysis in solution the formation of t^- is the rate-limiting step, while in the hydrolysis of esters the rate-limiting step occurs after the formation of t^-. In the case of amide hydrolysis by trypsin it is

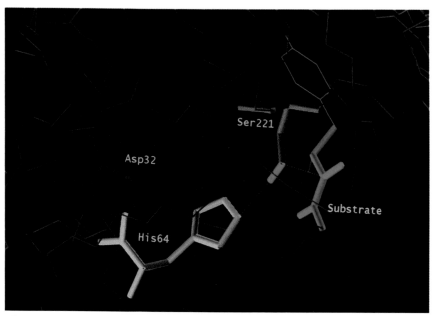

FIGURE 7.1. The active site of subtilisin. The residues of the catalytic triad (Asp 32, His 64 and Ser 221 are frequently denoted by the numbers of the corresponding residues in chymotrypsin (102, 57 and 195, respectively).

commonly assumed that the rate-limiting step is the formation of t^- and this will also be our working hypothesis. Mechanism b is referred to as the *charge-relay mechanism* or the *double-proton transfer mechanism* and is presented in many text books that discuss enzyme mechanism. This mechanism requires that the proton transfer from Ser_c to His_c will be accompanied by a concerted proton transfer from His_c to Asp_c. Our analysis begins with mechanism a and is followed by a comparative study of mechanism b.

7.2. POTENTIAL SURFACES FOR AMIDE HYDROLYSIS IN SOLUTION AND IN SERINE PROTEASES

7.2.1. The Key Resonance Structures for the Hydrolysis Reaction

In order to explore mechanism a, or any other mechanism, we have to start by defining the most important resonance structures and calibrating their energies using the relevant experimental information for the reference system in solution. The key resonance structures for the formation of t^- in mechanism a are

$$\psi_1 = Im \quad H\text{--}O \ C{=}O$$
$$\psi_2 = Im^+\text{---}H \ O^-C{=}O$$
$$\psi_3 = Im^+\text{---}H \ O\text{--}C\text{--}O^- \qquad (7.4)$$

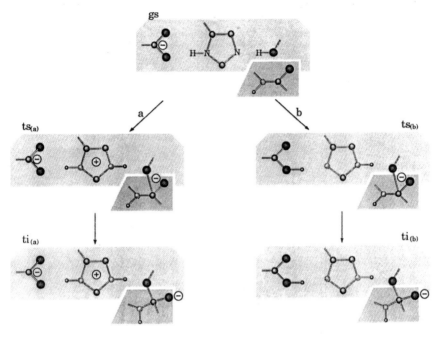

FIGURE 7.2. Two alternative mechanisms for the catalytic reaction of serine proteases. Route *a* corresponds to the electrostatic catalysis mechanism while route *b* corresponds to the double proton transfer (or the charge relay mechanism). gs ts and ti denote ground state, transition state and tetrahedral intermediate, respectively.

where Im, H–O, and C=O indicate, respectively, His_c, Ser_c, and the carbonyl of the substrate.

As before, we have to determine the energies associated with these resonance structures (i.e., the diagonal matrix elements). This is done conveniently using the functional forms suggested by the corresponding bonding configurations (see Fig. 7.3) and writing the EVB matrix elements in the all-atom solvent model as:

$$\varepsilon_1 = \Delta M(b_1) + \Delta M(b_7) + U_{nb}^{(1)} + U_{QQ}^{(1)} + U_{strain}^{(1)} + U_{Ss}^{(1)} + U_{ss}$$

$$\varepsilon_2 = \Delta M(b_3) + \Delta M(b_7) + U_{nb}^{(2)} + U_{QQ}^{(2)} + \alpha_2^0 + U_{strain}^{(2)} + U_{Ss}^{(2)} + U_{ss}$$

$$\varepsilon_3 = \Delta M(b_3) + \Delta M(b_4) + U_{nb}^{(3)} + U_{QQ}^{(3)} + \alpha_3^0 + U_{strain}^{(3)} + U_{Ss}^{(3)} + U_{ss}$$

$$H_{ij} = H_{ij}^0 = A_{ij} \exp\{-\mu(r_{kl(ij)} - r_{ij}^0)\} \qquad (7.5)$$

where the notation is the same as that used in eq. (6.4) and the relevant bonds, as well as the key energy terms, are given in Fig. 7.3. As in the previous case, the most important step is the calibration of the α_i^0 and the

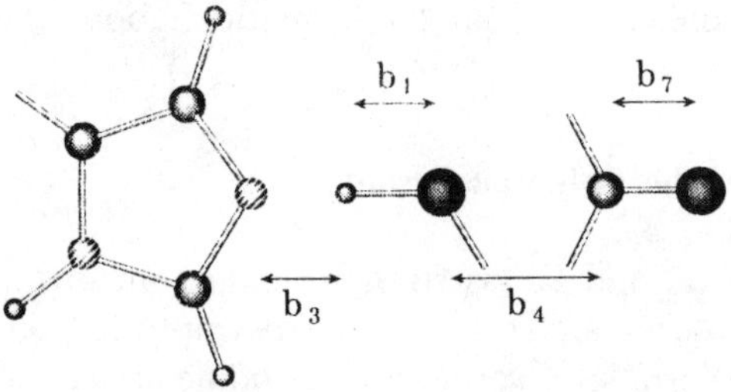

$$\epsilon_1 = \Delta M(b_1) + \Delta M(b_7) + \frac{1}{2}\sum_m K_b^{(1)}(b_m^{(1)} - b_0)^2 + \frac{1}{2}\sum_m K_\theta^{(1)}(\theta_m^{(1)} - \theta_0)^2 + K_\chi(\chi - \chi_0)^2$$

$$+ u_{QQ}^{(1)} + u_{nb}^{(1)} + u_{Qq,Ss}^{(1)} + u_{nb,Ss}^{(1)} + u_{ind,Ss}^{(1)} + u_{ss}$$

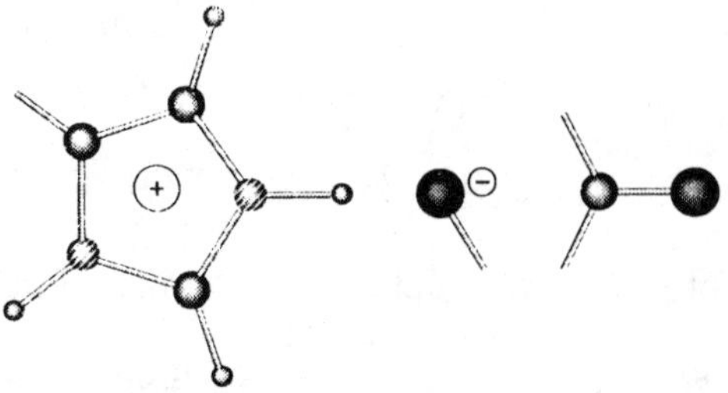

$$\epsilon_2 = \Delta M(b_3) + \Delta M(b_7) + \frac{1}{2}\sum_m K_b^{(2)}(b_m^{(2)} - b_0)^2 + \frac{1}{2}\sum_m K_\theta^{(2)}(\theta_m^{(2)} - \theta_0)^2 + K_\chi(\chi - \chi_0)^2$$

$$+ u_{QQ}^{(2)} + u_{nb}^{(2)} + \alpha^{(2)} + u_{Qq,Ss}^{(2)} + u_{nb,Ss}^{(2)} + u_{ind,Ss}^{(2)} + u_{ss}$$

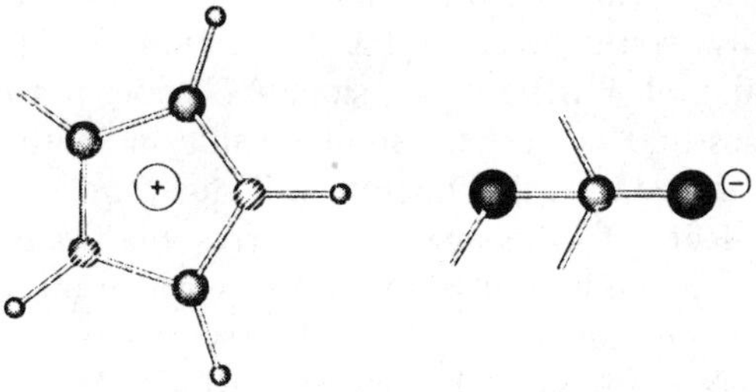

$$\epsilon_3 = \Delta M(b_3) + \Delta M(b_4) + \frac{1}{2}\sum_m K_b^{(3)}(b_m^{(3)} - b_0)^2 + \frac{1}{2}\sum_m K_\theta^{(3)}(\theta_m^{(3)} - \theta_0)^2$$

$$+ u_{QQ}^{(3)} + u_{nb}^{(3)} + \alpha^{(3)} + u_{Qq,Ss}^{(3)} + u_{nb,Ss}^{(3)} + u_{ind,Ss}^{(3)} + u_{ss}$$

FIGURE 7.3. The force fields for the three resonance structures that describe mechanism *a* for the catalytic reaction of serine proteases.

A_{ij} to reproduce the experimental information about the reaction in solution.

7.2.2. Calibrating the Potential Surface

The calibration of α_2^0 and H_{12} is straightforward since ψ_1 and ψ_2 describe a proton transfer process and the relevant asymptotic points are easily determined using the pK_a's of serine and histidine in water (see Chapter 5). The calibration of α_3^0 and A_{23} are more involved and require some effort in analyzing the available experimental information about $\Delta G_{2\rightarrow3}^{\infty}$ and $\Delta g_{2\rightarrow3}^{\neq}$ in water, which are considered below.

The value of $(\Delta G_{2\rightarrow3}^{\infty})_w$ can be obtained by writing

$$(\Delta G_{2\rightarrow3}^{\infty}) = \Delta G(R-O^- + C=O \rightarrow R-O-C-O^-)$$

$$= \Delta G_a(O^- + C=O + H^+ \rightarrow O-H+C=O)$$

$$+ \Delta G_b(O-H+C=O \rightarrow O-C-O-H)$$

$$+ \Delta G_c(O-C-O-H \rightarrow O-C-O^- + H^+) \quad (7.6)$$

The evaluation of ΔG_a, ΔG_b, and ΔG_c is considered in the exercise below.

Exercise 7.1. Estimate $(\Delta G_{2\rightarrow3}^{\infty})_w$ using only bond energies and pK_a values.

Solution 7.1. The value of ΔG_b can be estimated by noting that the relevant process involves a conversion of a C=O bond to two C–O bonds. The corresponding bond energies are 172 kcal/mol and 92 kcal/mol for the C=O and C–O bonds, respectively, giving $\Delta G_b \approx \Delta H_b \approx (-92\times 2) - (-172) = -12$ kcal/mol. A more reliable estimate can be obtained using group contributions (Ref. 3), which take into account the fact that the C=O bond is partially conjugated to the C=N bond. This correction gives $\Delta H_b \approx -0.5$ kcal/mol. Furthermore, since ΔG_b does not involve any charge transfer processes and has a very similar value in solutions and in the gas phase, one can use standard semiempirical quantum mechanical computer programs (e.g., Ref. 4) to estimate the corresponding ΔH_b. The values of ΔG_a and ΔG_c are much harder to obtain from quantum mechanical calculations but fortunately can be easily and very reliably obtained from pK_a values. That is, ΔG_a involves the process $(RO^- + H^+ \rightarrow R - OH)$ in solution and ΔG_c involves the process $(O-C-OH \rightarrow O-C-O^- + H^+)$. Thus we obtain (Ref. 5)

$$(\Delta G_{2\rightarrow3}^{\infty}) \approx \Delta H(R-OH+C=O \rightarrow R-O-C-OH)$$

$$- 2.3RT[pK_a(R-OH) - pK_a(O-C-OH)]$$

$$\approx -1 \text{ kcal/mol} \quad (7.7)$$

As demonstrated in the exercise above one can estimate the free energy of quite complicated processes by using bond energies and pK_a values.

The value of $(\Delta g_{2\to 3}^{\neq})_w^{\infty}$ can be estimated from experimental studies of methoxy-catalyzed hydrolysis of amides. That is, after some literature search you may find (Ref. 6) that the rate constant for an attack of $CH_3 - O^-$ on an amide is around $0.3\,\text{sec}^{-1}$. The corresponding $\Delta g^{\neq}$ is found in the exercise below.

Exercise 7.2. Find $(\Delta g_{2\to 3}^{\neq})_w^{\infty}$ by using the information given above about the corresponding rate constant (Hint: use some of the equations given in Chapter 2).

Solution 7.2. Using $k = 0.3\,\text{sec}^{-1}$, eq. (3.31) and eq. (2.12) will give $(\Delta g_{2\to 3}^{\neq})_w^{\infty} \simeq 17\,\text{kcal/mol}$. This value of $(\Delta g_{2\to 3}^{\neq})_w^{\infty}$ is expected to be reduced by $\sim 2\,\text{kcal/mol}$ when the ionized Im H^+ is brought near the $O^{-\delta} - C - O^{-\delta}$ transition state.

The above results give the asymptotic points of the potential surface in solution. Furthermore, with the use of the calculated solvation energies of the different fragments we can obtain from eq. (2.34) the asymptotic points for the gas-phase potential surface. This is done in Table 7.2.

Exercise 7.3. The discussion above gave you all the relevant information about the solution potential surface. Summarize this information in an energy diagram.

Solution 7.3. The corresponding diagram is given in Fig. 7.4.

With the estimates of Fig. 7.4 we can now determine α_3^0 and A_{23} by fitting the calculated surface for the $2 \to 3$ reaction in solution with Im H^+ at infinite distance, to the estimates of $(\Delta g_{2\to 3}^{\neq})_w$ and $(\Delta G_{2\to 3})_w$. This is done in Fig. 7.5. The parameters obtained in this way for H_{12} and the diagonal matrix elements are given in Table 7.3.

TABLE 7.2. Asymptotic Energy Values for the Reference Reaction in Solution and in the Gas Phase[a]

Resonance Forms		Notation	$(\Delta G_{i,w}^{\infty})_{\text{obs}}$	$\Delta g_{\text{sol},w}^{i,\infty}$	$\Delta G_{i,\text{gas}}^{\infty}$
Im	H–O C=O	ψ_1	0	−20	0
Im$^+$—H	O$^-$—C=O	ψ_2	12	−162	154
Im$^+$—H	O —C–O$^-$	ψ_3	11	−149	140

[a] The gas phase energies are estimated from the corresponding $(\Delta G_{1\to i})_w$, using $(\Delta G_{1\to i}^{\infty})_{\text{gas}} \simeq (\Delta G_{1\to i}^{\infty})_w - (\Delta g_{\text{sol},w}^{i,\infty} - \Delta g_{\text{sol},w}^{1,\infty})$, where the $\Delta g_{\text{sol},w}^{i,\infty}$ are estimated by eq. (2.34b) from the solvation energies of the relevant isolated fragments.

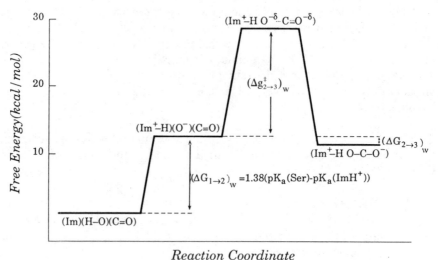

FIGURE 7.4. The energetics of the catalytic reaction of serine proteases in a reference solvent cage.

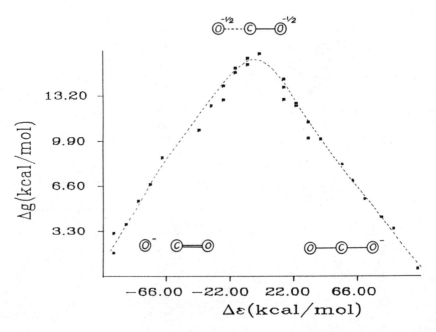

FIGURE 7.5. Calculated free-energy surface for the $2 \rightarrow 3$ step in solution. Forcing this surface to reproduce the observed value of $(\Delta g_{2 \rightarrow 3}^{\neq})_w^{\infty}$ is used to determine H_{23}.

TABLE 7.3. Parameters for the EVB Surface of Amide Hydolysis[a]

Bonds	$\Delta M = D[1 - \exp\{-a(b - b_0)\}]^2$		
O–H	$D = 102$	$b_0 = 0.96$	$a = 2.35$
N–H	$D = 93$	$b_0 = 1.00$	$a = 2.35$
N^+–H	$D = 93$	$b_0 = 1.00$	$a = 2.35$
C–O	$D = 92$	$b_0 = 1.43$	$a = 2.06$

Bond Angles	$U_\theta = \frac{1}{2}K_\theta(\theta - \theta_0)^2$	
X–C=O(ψ_1, ψ_2)	$K_\theta = 100$	$\theta_0 = 2.094$
X–C–O$^-$ (ψ_3)	$K_\theta = 100$	$\theta_0 = 1.911$
X–N^+–H(ψ_2, ψ_3)	$K_\theta = 100$	$\theta_0 = 2.094$

Nonbonded	$U_{nb} = A \exp\{-ar\}$	
O$\cdots$H	$A = 5288$	$a = 2.5$
O$^-\cdots$H	$A = 65$	$a = 2.5$
N$\cdots$H	$A = 150$	$a = 2.5$
N$\cdots$O$^-$		$a = 2.5$
N$\cdots$C		$a = 2.5$

Nonbonded[b]	$U'_{nb} = A_i A_j r^{-12} - B_i B_j r^{-6}$	
H	$A = 4$	$B = 0$
C	$A = 632$	$B = 20$
C^+	$A = 632$	$B = 20$
O	$A = 774$	$B = 24$
O$^-$	$A = 1140$	$B = 24$
N	$A = 774$	$B = 24$

Charges	$U_{qq} = 332q_i q_j/r_{ij}$		
$(O_a - H)(\psi_1)$	$q_{O_u} = -0.4$		$q_H = 0.4$
$(O_a^-)(\psi_2)$	$q_{O_a} = -1.0$		
$(C = O_b)(\psi_1, \psi_2)$	$q_C = 0.3$		$q_{O_b} = -0.3$
$(O_a - C - O_b^-)(\psi_3)$	$q_{O_a} = -0.2$	$q_C = 0.2$	$q_{O_b} = -1.0$
Im(ψ_1)	Taken from Fig. 5.4		
Im $H^+(\psi_2, \psi_3)$	Taken from Fig. 5.4		

Off-Diagonal Parameters and Diagonal α's			
H_{12}	$A_{12}^{NO} = -140$	$r^0 = 0.0$	$\mu_{12}^{NO} = 0.8$
H_{13}	$A = 0$	$r^0 = 0.0$	$\mu = 0$
H_{23}	$A_{23}^{O_a O_b} = -120$	$r^0 = 0.0$	$\mu_{23}^{O_a O_b} = 0.4$
α_1	0.0		
α_2	120		
α_3	126		

[a] Energies in kcal/mol, distance in Å, angles in radians, and charges in au. Parameters not listed in the table can be taken from Table 4.2.
[b] The function U_{nb} is used for the nonbonded interaction between the indicated EVB atoms while U'_{nb} is used for the nonbonded interactions between other atoms which are not bonded to each other or to a common atom.

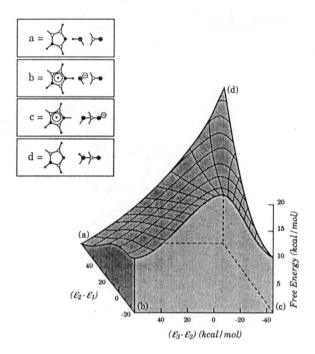

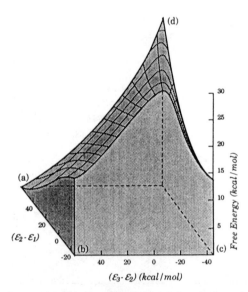

FIGURE 7.6. Comparing the potential surfaces for the catalytic reaction of trypsin (upper figure) to the corresponding reaction in solution (lower figure). The different configurations that define the corners of the potential surface are drawn on the upper left portion of the figure.

With the parameters of Table 7.3 and eq. (5.20) we can simulate the reaction in the enzyme-active site, replacing $(U^i_{Ss} + U_{ss})$ in eq. (7.5) by $(U^i_{Sp} + U_{pp})$ and comparing the resulting free-energy surface to the surface for the corresponding reaction in a reference solvent cage. Such a comparison is presented in Fig. 7.6. As seen from the figure, the enzyme appears to stabilize the transition state more than water does. The reason for this stabilization is apparent from Fig. 7.7; that is, the enzyme creates a network of oriented dipoles around the $(-+-)$ configuration of the transition state. This network involves two hydrogen bonds near the carbonyl carbon (which are called the *oxyanion hole* and stabilize the $-C-O^-$ oxyanion intermediate) and three dipoles near Asp 102 (which we will call the *Asp hole*). This situation is not much different from the one in the active site of lysozyme (Fig. 6.11).

Exercise 7.4. (a) Use the parameters of Table 7.3 and the LD model to calculate the activation energy of the $2 \rightarrow 3$ step in solution. (b) Repeat the same calculation in a protein model where a positive charge of $+0.5$ (3 Å from the carbonyl carbon) represents the oxyanion holes, while a negative charge of -0.5 near the His$^+$ residue represents the somewhat screened Asp 102. Simulate the rest of the system by the LD model.

Solution 7.4 Use Program 2.B.

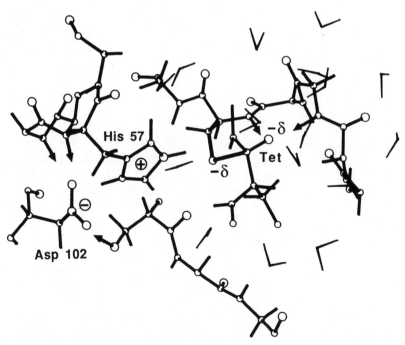

FIGURE 7.7. The protein dipoles (hydrogen bonds) that stabilize the $(-+-)$ transition state of trypsin.

7.3. EXAMINING THE CHARGE-RELAY MECHANISM

The considerations presented above were based on the specific assumption that the catalytic reaction of the serine proteases involves mechanism a of Fig. 7.2. However, one can argue that the relevant mechanism is mechanism b (the so-called "charge-relay mechanism"). In principle the proper procedure, in case of uncertainty about the actual mechanism, is to perform the calculations for the different alternative mechanisms and to find out which of the calculated activation barriers reproduces the observed one. This procedure, however, can be used with confidence only if the calculations are sufficiently reliable. Fortunately, in many cases one can judge the feasibility of different mechanisms without fully quantitative calculations by a simple conceptual consideration based on the EVB philosophy. To see this point let us consider the feasibility of the charge-relay mechanism (mechanism b) as an alternative to mechanism a. Starting from Fig. 7.2 we note that the energetics of route b can be obtained from the difference between the activation barriers of route b and route a by

$$\Delta g_b^{\neq} = \Delta g_a^{\neq} + \Delta\Delta g_{a \to b}^{\neq} \qquad (7.8)$$

If $\Delta\Delta g_{a \to b}^{\neq}$ is positive, then route b is practically blocked. As seen from Fig. 7.2, $\Delta\Delta g_{a \to b}^{\neq}$ is basically the free energy associated with a proton transfer from His_c to Asp_c at the transition state. This free energy can be evaluated in two steps. First, we estimate the free energy for this process in water and then evaluate the change in free energy upon transfer of the reacting fragments from water to the protein active site. The energetics in water is estimated in Fig. 7.8a and in Exercise 7.5.

Exercise 7.5. Estimate the free-energy difference between $(ts)_a$ and $(ts)_b$ (Fig. 7.8) in water.

Solution 7.5. The relevant thermodynamic cycle involves the electrostatic work of taking the Asp_c^- His_c^+ t^- system from the initial configuration in the solvent cage to infinity, the free energy of proton transfer from His to Asp at infinite separation, and the electrostatic work of returning the Asp_c His_c neutral pair and t^- to the same solvent cage. The free energy for the proton transfer process, ΔG_{PT}^{∞}, can be evaluated easily using the pK_a's of Asp and His in water. This gives

$$\Delta G_1 = \Delta G_{PT,w}^{\infty} = 1.38(pK_a(\text{Im H}^+) - pK_a(\text{Asp})) \approx 4.5 \, \text{kcal/mol} \qquad (7.9)$$

The electrostatic free energy associated with the separation of the ion pair and the recombination of the neutral pair can be easily calculated with Coulomb's law and a large dielectric constant. (e.g., $\varepsilon = 40$, which is the

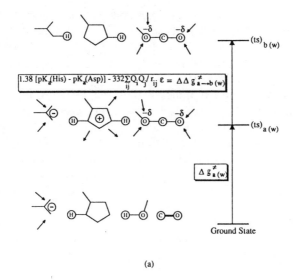

(a)

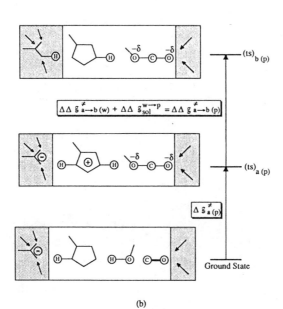

(b)

FIGURE 7.8. Comparing the energies of the transition states for mechanism a and b in solution (upper figure) and in the enzyme-active site (lower figure).

lower limit for ε in water when the two charges are in a close proximity) giving

$$\Delta G_2 = -V_{QQ}/\varepsilon \simeq -332 \sum_{ij} Q_i Q_j / r_{ij} 40 \simeq 2 \tag{7.10}$$

where V_{QQ} is the interaction between the residual charge on the given fragments (this energy can also be estimated by representing Asp_c^-, His_c^+, and t^- by point charges). The total energy of this process is now $(\Delta g_{a \to b}^{\neq})_w = \Delta G_1 + \Delta G_2 \simeq 7 \, kcal/mol$.

The results given above indicate that the charge-relay mechanism is unfavorable in water. This finding is also supported by experimental studies with model compounds (Ref. 7). One may still argue that the protein might make $\Delta \Delta g_{a \to b}^{\neq}$ negative. This question, however, should not be left as a major open hypothesis since it can be easily examined by PDLD calculations of the energetics associated with moving the transition states of a and b from the solvent cage to the protein-active site. Such a calculation yields an increase of $\Delta \Delta g_{a \to b}^{\neq}$ by an additional 6 kcal/mol, giving a total value of 12 kcal/mol for $\Delta \Delta g_{a \to b}^{\neq}$ (see Fig. 7.8b).

To realize the reason for this result from a simple intuitive point of view it is important to recognize that the ionized form of Asp_c is more stable in the protein-active site than in water, due to its stabilization by three hydrogen bonds (Fig. 7.7). This point is clear from the fact that the observed pK_a of the acid is around 3 in chymotrypsin, while it is around 4 in solution. As the stability of the negative charge on Asp_c increases, the propensity for a proton transfer from His_c to Asp_c decreases.

These points are also supported by additional experimental information. That is, neutron diffraction experiments (Ref. 8) on a complex of the inhibitor monoisopropylphosphoryl (MIP) and trypsin located on His_c the proton that bridges Asp_c and His_c (forming an Asp_c^- His_c^+ pair). This finding is relevant to the situation at the transition state since the inhibited MIP involves a negatively charged PO_3^- group at the site occupied by the oxyanion intermediate (although the difference in charge distribution between the two prevents one from reaching a unique conclusion).

7.4. SITE-SPECIFIC MUTATIONS PROVIDE A POWERFUL WAY OF EXPLORING DIFFERENT CATALYTIC MECHANISMS

The family of serine proteases has been subjected to intensive studies of site-directed mutagenesis. These experiments provide unique information about the contributions of individual amino acids to k_{cat} and K_M. Some of the clearest conclusions have emerged from studies in subtilisin (Ref. 9), where the oxyanion intermediate is stabilized by the main-chain hydrogen bond of Ser 221 and an hydrogen bond from Asn 155 (Ref. 2). Replacement of Asn 155 (e.g., the Asn 155 → Ala 155 described in Fig. 7.9) allows for a quantitative assessment of the effect of the protein dipoles on $\Delta g^{\neq}$.

The FEP and PDLD approaches developed in the previous chapters can be used conveniently to calculate the effect of genetic mutations. For example, one can calculate the reaction profile for the native and mutant

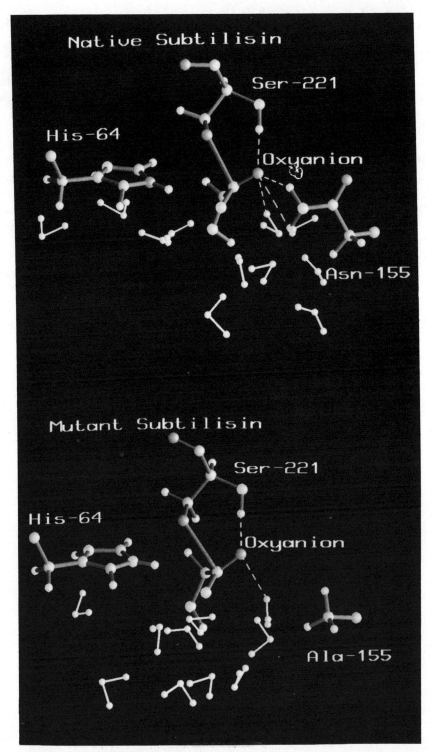

FIGURE 7.9. The Asn-155→Ala mutation in subtilisin involves deletion of a hydrogen bond between the enzyme and the oxyanion transition state.

enzyme and evaluate the difference in $\Delta g^{\neq}$ ($\Delta\Delta g^{\neq}$) associated with the mutation. Such a thermodynamic cycle [which is denoted in Fig. 7.10 by ($\Delta G_2' - \Delta G_2$)] can be considered formally as a "mutation" of the substrate between its ground state and transition state, in the native and mutant enzymes. This type of calculation will give, as a byproduct, the location of the transition states in the native and mutant enzymes. Once the transition states are located we can try an alternative thermodynamic cycle, mutating the protein at the (ES) and (ES$^{\neq}$) states rather than "mutating" the substrate from its ground to transition state at the native and mutant enzyme (the $\Delta G_c - \Delta G_b$ cycle of Fig. 7.10). Similarly one can calculate the effect of mutations on binding free energy (the ΔG_s of Fig. 5.2) in an *indirect* way, mutating the protein at the E + S and ES states and obtaining $\Delta\Delta G_{\mathrm{bind}}$ from the $\Delta G_a - \Delta G_b$ of Fig. 7.10.

For what is probably the earliest microscopic calculations of thermodynamic cycles in proteins see Ref. 12, that reported a PDLD study of the pK_a's of some groups in lysozyme. The use of FEP approaches for studies of proteins is more recent and early studies of catalysis and binding were reported in Refs. 11, 12, and 13 of Chapter 4.

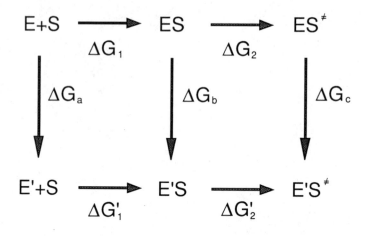

$$\Delta\Delta G_{\mathrm{bind}} = \Delta G_1 - \Delta G_1' = \Delta G_a - \Delta G_b$$

$$\Delta\Delta g_{\mathrm{cat}}^{\neq} = \Delta G_2' - \Delta G_2 = \Delta G_c - \Delta G_b$$

FIGURE 7.10. Different thermodynamic cycles that can be used to determine the effect of mutations on activation-free energies and binding-free energies. The figure designates the native and mutant enzymes by E and E', respectively. Note that one can either mutate the substrate between the ground and transition state or mutate the proteins at the ground and transition state (this, however, requires one to find the location of the transition state).

The results of a study of several mutation of Asn 155 in subtilisin are presented in Fig. 7.11. The agreement between the calculated (Ref. 5b) and observed (Ref. 9) results is almost quantitative, providing a powerful verification of structure–function correlations against a clear data base (which does not involve some of the uncertainties associated with comparison of enzymatic reactions to the corresponding reactions in solutions). Moreover, calculations of the effects of point mutations offer much more than the verification of the given theoretical approach. That is, while genetic substitution tells us what is the contribution of a given group to $\Delta g^{\neq}$, it does not tell us in a direct way what are the energy components of the given contribution. For example, the substitution of Asp_c in subtilisin leads to a change of 4.6 kcal/mol in $\Delta g^{\neq}$ (Ref. 10a) and a similar effect is observed in trypsin (Ref. 10b). It is not clear, however, whether this is due to elimination of the charge relay mechanism or to the loss of the electrostatic

Enzymes	$\Delta\Delta G_{bind}$		$\Delta\Delta g^{\neq}_{cat}$	
	exp.	calc.	exp.	calc.
‒ ‒ ‒ ‒ ‒ Thr			4.7	5.0
⋯⋯⋯ Ala	-0.6	-0.2	3.7	4.1
‒ ⋅ ‒ ⋅ ‒ ⋅⋅ Leu	~0.	0.5	3.3	3.7
——— Asn	0.	0.	0.	0.

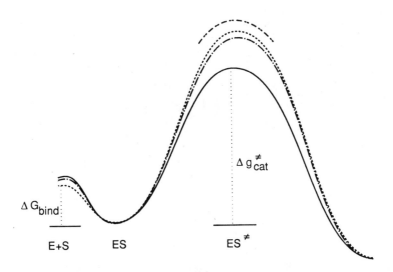

FIGURE 7.11. Calculated and observed free-energy changes for the Asn-155→Thr, the Asn-155→Leu, and the Asn-155→Ala mutations. The calculated and observed free energies are compared in the table in the upper part of the figure.

stabilization of His_c^+ by Asp_c^- in mechanism a of Fig. 7.2. Here one can calculate the actual contributions to $\Delta\Delta g^{\neq}$ and analyze their relative magnitude, under the constraint that the total calculated change in $\Delta g^{\neq}$ should reproduce the corresponding observed value (Ref. 11). Calculations which are capable of reproducing the observed $\Delta\Delta g^{\neq}$ in an extensive number of test cases are probably sufficiently reliable to tell us which mechanism is responsible for the given catalytic effect.

REFERENCES

1. D. M. Blow, J. J. Briktoft, and B. S. Hartley, *Nature*, **221**, 337 (1969).

2. J. Kraut, *Ann. Rev. Biochem.*, **46**, 331 (1977).

3. S. W. Benson, *Thermochemical Kinetics*, Wiley, New York, 1968.

4. J. J. P. Stewart, QCPE No. 455, Indiana University, 1986.

5. (a) A. Warshel and S. Russell, *J. Am. Chem. Soc.*, **108**, 6569 (1986). (b) A. Warshel, F. Sussman, and J-K. Hwang, *J. Mol. Biol.*, **201**, 139 (1988).

6. A. R. Fersht, *J. Am. Chem. Soc.*, **93**, 3504 (1971).

7. G. A. Rogers and T. C. Bruice, *J. Am. Chem. Soc.*, **96**, 2473 (1974).

8. A. A. Kossiakoff and S. A. Spencer, *Biochemistry*, **20**, 6462 (1981).

9. (a) J. A. Wells, B. C. Cunningham, T. P. Graycar, and D. A. Estell, *Phil. Trans. R. Soc. London*, No. 317, 415 (1986). (b) P. Bryan, M. W. Pontoliano, S. G. Quill, H. Y. Hsiao, and T. Poulos, *Proc. Natl. Acad. Sci. U.S.A.*, **83**, 3743 (1986).

10. (a) P. Carter and J. A. Wells, *Nature*, **332**, 564 (1988). (b) C. S. Craik, S. Roczniak, C. Largeman, and W. J. Rutter, *Science*, **237**, 909 (1987).

11. A. Warshel, G. Naray-Szabo, F. Sussman, and J-K. Hwang, *Biochemistry*, **28**, 3629 (1986).

12. A. Warshel, *Biochemistry*, **20**, 3167 (1981).

8

SIMULATING METALLOENZYMES

8.1. STAPHYLOCOCCAL NUCLEASE

8.1.1. The Reaction Mechanism and the Relevant Resonance Structures

Staphylococcal nuclease (SNase) is a single-peptide chain enzyme consisting of 149 amino acid residues. It catalyzes the hydrolysis of both DNA and RNA at the 5′ position of the phosphodiester bond, yielding a free 5′-hydroxyl group and a 3′-phosphate monoester

$$H_2O + 5' - OP(O_2)^- O - 3' \rightleftharpoons 5' - OH + (OH)P(O_2)^- O - 3' \quad (8.1)$$

The enzyme requires one Ca^{2+} ion for its action and shows little or no activity when Ca^{2+} is replaced by other divalent cations. A crystallographic structure at 1.5 Å resolution of SNase in complex with the inhibitor pdTp has been determined by Cotton and co-workers (Ref. 1). The active site (Fig. 8.1) is located at the surface of the protein with the pyrimidine ring of pdTp fitting into a hydrophobic pocket while the 3′- and 5′-phosphate groups interact with several charged groups. In particular, the two arginine residues, 35 and 87, donate hydrogen bonds to the 5′-phosphate, thereby partly neutralizing its double negative charge. The Ca^{2+} ion is ligated by the

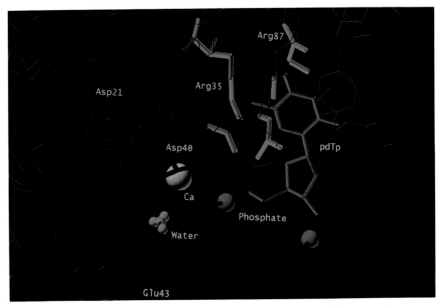

FIGURE 8.1. The structure of the active site of SNase with a bound inhibitor that is used as a model for the substrate.

carboxylate groups of Asp 21 and Asp 40, the carbonyl oxygen of Thr 41, two water molecules, and one of the 5'-phosphate oxygens.

Based on this protein-inhibitor structure, a reaction mechanism for the enzyme has been postulated (Ref. 1): (1) general base catalysis by Glu 43, which accepts a proton from a (crystallographically observed) water molecule in the second ligand sphere of the Ca^{2+} ion, yielding a free hydroxide ion; (2) nucleophilic attack by the OH^- ion on the phosphorus atom in line with the 5'–O–P ester bond, leading to the formation of a trigonal bipyramidal (i.e., pentacoordinated) transition state or metastable intermediate; (3) breakage of the 5'–O–P bond and formation of products.

The overall catalytic rate constant of SNase is (see, for example, Ref. 3) $k_{cat} \simeq 95 s^{-1}$ at $T = 297$ K, corresponding to a total free energy barrier of $\Delta g^{\ddagger}_{cat} = 14.9$ kcal/mol. This should be compared to the pseudo-first-order rate constant for nonenzymatic hydrolysis of a phosphodiester bond (with a water molecule as the attacking nucleophile) which is $2 \times 10^{-14} s^{-1}$, corresponding to $\Delta g^{\ddagger}_w = 36$ kcal/mol. The rate increase accomplished by the enzyme is thus 10^{15}–10^{16}, which is quite impressive.

The first two steps of the SNase reaction, of which the second one is rate limiting, can be described by the three EVB resonance structures of Fig. 8.2. Here, ψ^p_1 represents the reactant state, with Glu 43 negatively charged and the 5'-phosphate group in tetrahedral conformation. The state resulting from the general base catalysis step, where Glu 43 has been protonated by the adjacent water molecule, is denoted by ψ^p_2, and the state with the pentacoordinated phosphate group formed after nucleophilic attack by the

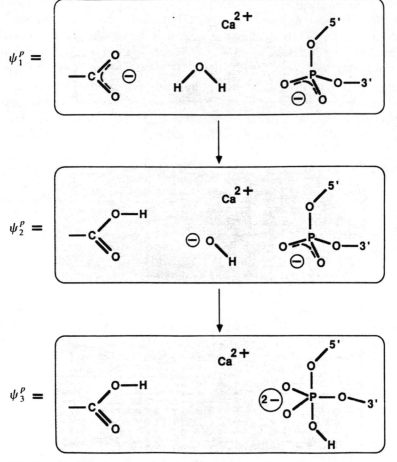

FIGURE 8.2. The resonance structures for the proposed mechanism of SNase.

OH^- ion is denoted ψ_3^P. The atoms depicted in the figure are considered as our solute system (S) while the rest of the protein–water environment constitutes the "solvent" (s) for the enzyme reaction. Although the Ca^{2+} ion does not actually "react," it is included in the reacting system for convenience. As before, we describe the diagonal elements of the EVB Hamiltonian associated with the three resonance structures (ψ_1^P, ψ_2^P, ψ_3^P) by classical force fields, using:

$$\varepsilon_j = H_{jj} = \sum_m \Delta M_m^{(j)}(b_m^{(j)}) + \frac{1}{2}\sum_m K_b^{(j)}(b_m^{(j)} - b_{0,m}^{(j)})^2 + \frac{1}{2}\sum_m K_\theta^{(j)}(\theta_m^{(j)} - \theta_{0,m}^{(j)})^2$$

$$+ \sum_m K_{\phi,m}^{(j)}\cos(n_m^{(j)}\phi_m^{(j)}) + U_{QQ}^{(j)}$$

$$+ U_{nb}^{(j)} + U_{Qq,Ss}^{(j)} + U_{nb,Ss}^{(j)} + U_{ss} + \alpha_j^0 \qquad (8.2)$$

where, as in previous chapters, $\Delta M_m^{(j)}$ denotes the Morse potential (relative to its minimum value) corresponding to the mth bond in the jth resonance structure. Bonds which are not included in the EVB list are described by a quadratic potential (note that K_b is set to zero for the EVB bonds). The third and fourth terms are the bond-angle and dihedral-angle bending contributions. $U_{QQ}^{(j)}$ denotes the electrostatic interaction between the solute charges and $U_{nb}^{(j)}$ designates the solute nonbonded interaction (other than electrostatic). The interaction energy between the solute system and the surrounding protein–water is contained in $U_{Qq,Ss}^{(j)}$, the electrostatic part, and $U_{nb,Ss}^{(j)}$ the rest of the nonbonded interaction.

8.1.2. The Construction of the EVB Potential Surface for the Reaction

The determination of the $\Delta G_{i \to j}$'s depends, of course, on the choice of the reference reaction in solution. For instance, when one states that the rate enhancement by SNase is $\sim 10^{16}$, one makes the implicit assumption of the reference reaction being

$$H_2O + (CH_3O)_2PO_2^- \rightleftharpoons (CH_3O)_2P(OH)_2O^- \qquad (8.3)$$

where the attacking species is a water molecule (from now on we only consider the reactions up to the formation of the pentacoordinated intermediate–transition state since this is the rate-limiting step). The activation free-energy barrier for this reaction is 36 kcal/mol. This is, however, not the mechanism proposed for SNase, which involves an hydroxide ion as the attacking species. A more useful choice of reference reaction in solution would therefore be

$$OH^- + (CH_3O)_2PO_2^- \rightleftharpoons (CH_3O)_2P(OH)O_2^{2-} \qquad (8.4)$$

This reaction requires the formation of an hydroxide ion, as in the enzyme reaction. A proper reference reaction for the first step in the enzyme would then be simply the proton transfer from a water molecule to a glutamic acid in solution:

$$(Glu) - COO^- + H_2O \rightleftharpoons (Glu) - COOH + OH^- \qquad (8.5)$$

The observed reaction free energy for this step is given by $(\Delta G_{1 \to 2})_w = 2.3\,RT(pK_a[H_2O] - pK_a[Glu]) = 15.9\,kcal/mol$, while the activation free energy is estimated to be $(\Delta g_{1 \to 2}^\ddagger)_w \simeq 18.3\,kcal/mol$ at 297 K, using data from the reaction $H_2O \rightleftharpoons H^+ + OH^-$. The free energies and rate constants for formation of pentacoordinated intermediates for various phosphate ester hydrolysis reactions have been calculated and compiled by Guthrie (Ref. 2). For the hydrolysis of dimethylphosphate by OH^- [eq. 8.4)] the obtained values are $(\Delta G_{2 \to 3})_w = 22(\pm 3)\,kcal/mol$ and $(\Delta g_{2 \to 3}^\ddagger)_w = 33\,kcal/mol$. We thus have the reference free-energy diagram depicted in Fig. 8.3 from the experimental solution data. It should be noted

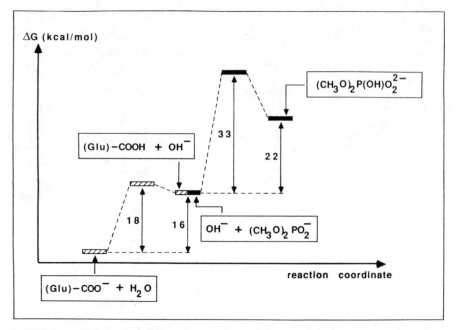

FIGURE 8.3. The energetics of an hypothetical reference reaction that corresponds to the assumed mechanism of SNase but occurs in a solvent cage.

that if the reaction proceeds through exactly the same mechanism in solution as in the enzyme (including the proton transfer to a glutamic acid), the total free-energy barrier will be almost 50 kcal/mol, corresponding to an enzyme rate acceleration of 10^{25}! However, our reference reaction corresponds to a convenient mathematical trick that guarantees a properly calibrated surface for the given enzymatic reaction and does not have to represent the actual mechanism in solution.

Now we are ready to calibrate our EVB surface for the solution reaction. To do this we start with the first step and consider the two resonance structures

$$\psi_1 = (O - C - O)^-(H - O - H)(PO_2^-(OR)_2)$$

$$\psi_2 = (O - C - OH)(OH)^-(PO_2^-(OR)_2) \tag{8.6}$$

The corresponding calibration process is given as an exercise below.

Exercise 8.1. Find α_1, α_2, and H_{12} for the proton transfer step by using the above experimental information and Program 2.B.

After performing this exercise you will get similar parameters to those obtained by more elaborated free-energy calculations and summarized in Table 8.1. A similar procedure can be used for the second step where the

TABLE 8.1. Parameters for the EVB Potential Surface of the Reaction of Staphylococcal Nuclease[a]

Bonds	$\Delta M(b) = D[1 - \exp\{-a(b - b_0)\}]^2$		
C=O (ψ_1, ψ_2, ψ_3)	$D = 120$	$b_0 = 1.25$	$a = 2.0$
C–O (ψ_2, ψ_3)	$D = 83$	$b_0 = 1.36$	$a = 2.0$
O–H (ψ_1, ψ_2, ψ_3)	$D = 109$	$b_0 = 1.00$	$a = 2.0$
P–O (ψ_1, ψ_2, ψ_3)	$D = 83$	$b_0 = 1.60$	$a = 2.0$
P=O (ψ_1, ψ_2)	$D = 120$	$b_0 = 1.49$	$a = 2.0$

Bond Angles	$U_\theta = \frac{1}{2}K_\theta(\theta - \theta_0)^2$	
O–P–O (ψ_1, ψ_2)	$K_\theta = 60$	$\theta_0 = 109°$
O–P–O (ψ_3)[b]	$K_\theta = 60$	$\theta_0 = 90°$
O–P–O (ψ_3)[b]	$K_\theta = 60$	$\theta_0 = 120°$
O–P–O (ψ_3)[b]	$K_\theta = 60$	$\theta_0 = 180°$

Nonbonded[c]	$U_{nb} = A_{ij}e^{-ar}$	
$O \cdots O$ (ψ_2)	$A = 3600$	$a = 2.5$
$O \cdots P$ (ψ_2)	$A = 3900$	$a = 2.5$

Nonbonded	$U'_{nb} = A_iA_jr^{-12} - B_iB_jr^{-6}$	
H	$A = 4$	$B = 0$
O	$A = 1120$	$B = 24$
C	$A = 632$	$B = 24$
P	$A = 1500$	$B = 24$
Ca	$A = 345$	$B = 15$

Charges	$U_{qq} = 332q_iq_j/r_{ij}$		
$(O–C–O)^-$ (ψ_1)	$q_O = -0.7$	$q_C = +0.4$	
H–O–C=O (ψ_2, ψ_3)	$q_H = +0.4$	$q_O = -0.4$	$q_C = +0.4$
H–O–H (ψ_1)	$q_H = +0.4$	$q_O = -0.8$	
$(H–O)^-$ (ψ_2)	$q_H = 0.0$	$q_O = -1.0$	
5'–O(HO)P(OO$^-$)O–3' (ψ_3)	$q_H = 0.0$	$q_{O_1} = q_{O_2} = q_{O_5} = -0.4$	$q_P = +1.0$
	$q_{O_3} = q_{O_4} = -0.9$		
5'–OP(OO$^-$)O–3' (ψ_1, ψ_2)	$q_{O_1} = q_{O_4} = -0.36$	$q_P = +0.99$	$q_{O_2} = q_{O_3} = -0.635$
Ca^{2+} (ψ_1, ψ_2, ψ_3)	$q_{Ca} = +2.0$		

Off-Diagonal Parameters and Diagonal Shifts			
H_{12}	$A_{12}^{OO} = 10$	$\mu_{12}^{OO} = 0.0$	$r_0 = 0.0$
H_{23}	$A_{23}^{OP} = 35$	$\mu_{23}^{OP} = 0.0$	$r_0 = 0.0$
α_1	0.0		
α_2	22		
α_3	207		

[a] Energies are in kcal/mol, distances in Å, and atomic charges in au. Parameters not listed in the table are the same as in previous chapters.
[b] The three different functions correspond to the three possible O–P–O angles around the pentacoordinated phosphate.
[c] The nonbonded interaction term used for the $OH^- \cdots PO_4^-$ interaction in the EVB calculation.

(OH$^-$) attacks the phosphate group, considering the two resonance structures ψ_2 and ψ_3 of Fig. 8.2 (without the Ca^{2+} ion). The corresponding parameters for ε_2, ε_3, and H_{23} are also given in Table 8.1 (see Ref. 4 for more details).

8.1.3. The Ca^{2+} Ion Provides Major Electrostatic Stabilization to the Two High-Energy Resonance Structures

After obtaining the EVB parameters for the reaction in solution we are ready to consider the protein reaction. Here there is one new major element not considered in the previous chapters—the interaction of the reaction system with the metal. This might require consideration of the actual bonding between the metal and these fragments. However, as a zero-order approximation one can describe these interactions in terms of atom–atom electrostatic and van der Waals interactions. The corresponding parameters can be determined by either fitting potential functions to quantum mechanical calculations or by adjusting parameters to reproduce experimental information about the energetics and structure of the solvent around the metal in aqueous solution. This approach is taken here and the corresponding parameters are given in Table 8.1 (see Ref. 4 for more details). Apparently, the main effect of the metal is in providing electrostatic stabilization to both OH$^-$ in ψ_2 and the additional negative charge on the phosphate in ψ_3. This results in a major reduction of the activation free energy of the reaction, as demonstrated in Fig. 8.4. In the first step of the reaction the enzyme utilizes the Ca^{2+} charge to stabilize the hydroxide ion in a very significant way (in solution the proton transfer step costs about

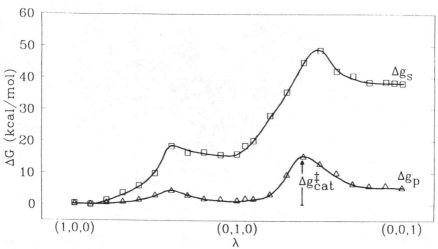

FIGURE 8.4. Calculated free-energy profiles for the reference reaction in solution, Δg_s, and for the enzyme reaction, Δg_p.

18 kcal/mol while the enzyme reduces the energetics of this step by almost 15 kcal/mol). In the second step the enzyme appears to work by providing an effective electrostatic complimentary to the transition state. That is, the loss of interaction energy between the Ca^{2+} ion and the hydroxide ion, in moving toward the pentacoordinated structure, is compensated for by increased interaction between the Ca^{2+} ion and the 5'-phosphate oxygen ligand. The accumulating negative charge $(-1 \rightarrow -2)$ on the phosphate

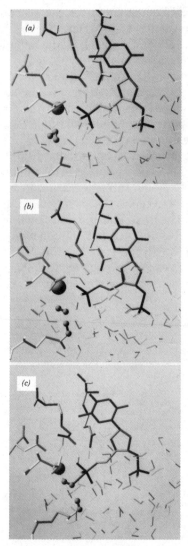

FIGURE 8.5. Three snapshots from the trajectories that lead from the ground state to the transition state in the catalytic reaction of SNase.

group is effectively stabilized by closer interactions with Arg 35 and Arg 87. In particular, Arg 87 appears to be an important factor, as its hydrogen bonds interact strongly with two of the phosphate oxygens in the transition state and not in the reactant state. This is also supported by the fact that a mutation of Arg 87 leads to a large effect on k_{cat} for this species.

Exercise 8.2. (a) Use the EVB Program 3.C and construct a potential surface for the reaction of Fig. 8.2, in the absence of the calcium ion, in water. (b) Examine the enzymatic reaction by adding the Ca^{2+} to the calculation of (a).

As emphasized in Chapter 5, we can use the analytical EVB potential surfaces to simulate the dynamics of our enzymatic reaction. This is done by propagating downhill trajectories from the different transition states, using the time reversal of these trajectories to construct the actual reactive trajectories (which are very rare and cannot be obtained by direct simulations). A few snapshots from our reactive trajectories are depicted in Fig. 8.5. The main point from this dynamical study, which requires more photographs for a clear illustration, is the fact that the Ca^{2+} ions helps the reaction by moving with the OH^- nucleophile toward the phosphate. (A movie of this reaction can be obtained from the author). This concerted motion allows the Ca^{2+} to retain the stabilization of the OH^- ion, while also helping the transfer of the OH^- charge to the phosphate oxygens (the Ca^{2+} also stabilizes the developing negative charge on the phosphate oxygens).

8.2. CARBONIC ANHYDRASE

The approach taken above estimates the effect of the metal by simply considering its electrostatic effect (subjected, of course, to the correct steric constraint as dictated by the metal van der Waals parameters). To examine the validity of this approach for other systems let's consider the reaction of the enzyme carbonic anhydrase, whose active site is shown in Fig. 8.6. The reaction of this enzyme involves the "hydration" of CO_2, which can be described as (Ref. 5)

$$Zn^{2+} + H_2O + CO_2 \rightleftharpoons Zn^{2+} \cdots OH^- + CO_2 + H^+ \rightleftharpoons Zn^{2+} \cdots HCO_3^- + H^+ \tag{8.7}$$

where the enzyme-active site uses a Zn^{2+} ion to catalyze the reaction. This reaction can be described by the VB structures considered in Exercise 8.3.

Exercise 8.3. Write the VB resonance structures for the reaction in eq. (8.7).

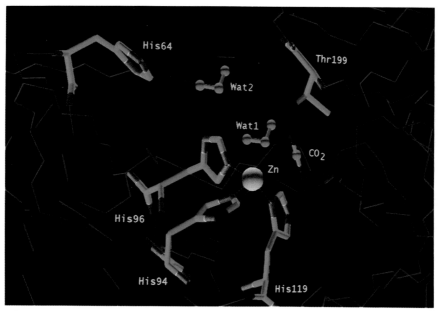

FIGURE 8.6. The catalytic site of carbonic anhydrase (Ref. 7). The water molecule is 2.2 Å from the Zn^{2+} ion and 2.6 Å from the carbon of the CO_2 which is held 2.5 Å from the Zn^{2+} ion.

Solution 8.3. This reaction can be described by

$$\psi_1 = H - O - H \quad O = C = O \, (Zn^{2+})$$

$$\psi_2 = H - O^- H^+ \quad O = C = O \, (Zn^{2+})$$

$$\psi_3 = H^+ \qquad H - O - \underset{\underset{O_-}{|}}{C} = O \, (Zn^{2+}) \tag{8.8}$$

where the H^+ ion is attached, of course, to a donor molecule (e.g. a water molecule).

With the valence bond structures of the exercise, we can try to estimate the effect of the enzyme just in terms of the *change* in the activation-free energy, correlating $\Delta\Delta g^{\neq}$ with the change in the electrostatic energy of ψ_2 and ψ_3 upon transfer from water to the enzyme-active site. To do this we must first analyze the energetics of the reaction in solution and this is the subject of the next exercise.

Exercise 8.4. Analyze the energetics of the CO_2 hydration reaction [eq. (8.7)] in solution.

Solution 8.4. To accomplish this task we have to find a simple cycle with easily available energies. Such a cycle is almost always available and indeed we note that the first step is a simple dissociation of water with pK_a of 15.7 and $\Delta G_1 = 21.4\,kcal/mol$. We also note that the second step can be described by the cycle

$$O=C=O + OH^- + H^+ \xrightarrow{\Delta G_{2,w}} {}^-O-\underset{\displaystyle O}{\overset{\displaystyle O}{\underset{\|}{C}}}-OH + H^+$$

$$\Bigg\downarrow \Delta G_{3,w} \qquad\qquad\qquad\qquad \Bigg\uparrow \Delta G_{5,w}$$

$$O=C=O + H_2O \xrightarrow{\Delta G_{4,w}} HO-\underset{\displaystyle O}{\overset{\displaystyle O}{\underset{\|}{C}}}-OH \qquad (8.9)$$

with $\Delta G_{4,w} = 0.1\,kcal/mol$ from the standard free energies of O=C=O, H_2O, and H_2CO_3 (−92.2, −56.6, and −148.7 kcal/mol, respectively), $\Delta G_3 = -21.4\,kcal/mol$ from the pK_a of water) and $\Delta G_{5,w} = 6.1\,kcal/mol$ from the pK_a of H_2CO_3 we obtain $\Delta G_{2,w} = \Delta G_{3,w} + \Delta G_{4,w} + \Delta G_{5,w} = -14.8\,kcal/mol$. Another estimate of $\Delta G_{2,w}$ can be obtained from the kinetic data of Ref. 6, which gives $k_{2\to3} \simeq 2.10^4\,s^{-1}$ and $k_{3\to2} \sim 2.10^{-4}$ s^{-1} (with the notation $CO_2 + OH^- \underset{k_{3\to2}}{\overset{k_{2\to3}}{\rightleftharpoons}} HCO_3^-$) which gives through eq. (2.3) $K_{2\to3} \sim 10^8$ and $(\Delta G_{2\to3})_w = -RT\ln K_{2\to3} \simeq -11\,kcal/mol$ [where $(\Delta G_{2\to3})_w$ is the $\Delta G_{2,w}$ of eq. (8.9)]. $(\Delta g^{\neq}_{2\to3})_w$ can be conveniently obtained from Ref. 5 using the value given above for $k_{2\to3}$, eqs. (3.31) and (2.12), which gives $(\Delta g^{\neq}_{2\to3})_w \simeq 11.5\,kcal/mol$. Thus we obtain the energetics depicted in Fig. 8.7.

Once the energetics of the reference reaction are estimated we are ready to analyse the effect of the enzyme, which reduces the barrier from $\sim 25\,kcal/mol$ to $\sim 9\,kcal/mol$, with the first step ($H_2O \to H^+ + OH^-$) as the

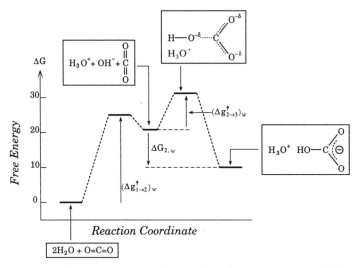

FIGURE 8.7. A schematic free-energy diagram for a stepwise hydration of CO_2 in water.

rate limiting step (Ref. 5). This analysis involves a minor complication since the transfer of the proton to the water molecule is followed by its transfer to an histidine residue and to solution, before the nucleophilic attack step. Thus the initial water splitting process should be considered as a two-step mechanism, which lowers the reference energy for the nucleophilic attack step. For this mechanism we will have to consider the pK_a difference between H_3O^+ and histidine. Nevertheless, for simplicity we suggest that the reader neglect the secondary proton transfer step and follow the exercise below but remember that the actual situation is somewhat more complicated (Ref. 16).

Exercise 8.5. Try to estimate the catalytic effect of carbonic anhydrase by evaluating the energetics of the reacting fragments in solution and in a simplified LD enzyme model with Zn^{2+} and three surrounding histidine residues. Use the geometry of Fig. 8.6 for the reacting system and ignore the secondary proton transfer step.

Solution 8.5. First, use the LD model to calculate the $\Delta g^i_{sol,w}$ [the results should be -25, -220, and -190 kcal/mol for $\Delta g^{1,\infty}_{sol,w}$, $\Delta g^{2,\infty}_{sol,w}$ and $\Delta g^{3,\infty}_{sol,w}$, respectively]. Now you should repeat the calculations, modeling the protein-active site that includes the Zn^{2+} ion as well as the other protein residues by the PDLD model.

The exercise given above should *overestimate* the activation barrier in the enzyme, since it does not take into account the secondary transfer of the proton from water to histidine. A more complete study (Fig. 8.8) that

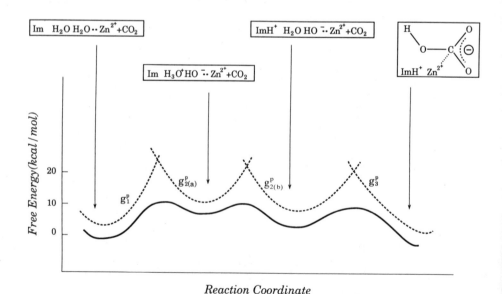

FIGURE 8.8. Calculated free-energy profile for the reaction of carbonic anhydrase. $g_{2(a)}$ and $g_{2(b)}$ designate the states where the proton acceptors are water and histidine respectively.

considers this transfer reproduces the actual catalytic activity of the enzyme (Ref. 16).

8.3. GENERAL ASPECTS OF METALLOENZYMES

8.3.1. Linear Free-Energy Relationships for Metal Substitution

The two examples given above indicate that the role of the metal ion can be captured by considering its electrostatic effect. This, however, must be done with care, taking into account the specific ionic radius of the metal and its van der Waals interactions with the nucleofile and the substrate. A useful way to analyze the trend associated with the metal size is to consider the effect of metal substitution in SNase. For simplicity we will consider first the effect of the metal radius on ψ_2 and ψ_3 and examine the effect on ψ_1 only in the final treatment. We will look for the trend in moving from a large ion (Ba^{2+}) to an intermediate ion (Ca^{2+}) and to a small ion (Mg^{2+}). In changing the ion size one may expect several basic types of "selectivity" patterns for the rate constant as a consequence of different dependence of the two states on the ion properties (see Ref. 8 for general considerations of ion selectivity). This is considered in Fig. 8.9, which depicts four limiting cases: in Fig. 8.9a, ψ_3 is less sensitive to the ion size than ψ_2 over the entire range of the ionic radius (r_{ion}) considered. Hence, the larger the ion, the higher the rate constant will be, $k(Ba^{2+}) > k(Ca^{2+}) > k(Mg^{2+})$. If, on the other hand, ψ_2 is less sensitive to the ion radius, we will obtain the opposite ordering between the rates, $k(Ba^{2+}) < k(Ca^{2+}) < k(Mg^{2+})$ (Fig. 8.9b). As a third case, one can imagine the possibility that ψ_2 is more sensitive to larger ions while ψ_3 is more sensitive to smaller ions. This case is depicted in Fig. 8.9c and would lead to a maximum of the activation barrier for the intermediate ion, $k(Ba^{2+}) > k(Ca^{2+}) < k(Mg^{2+})$. The only case which could give a minimum barrier for the intermediate ion is shown in Fig. 8.9d, in which the sensitivities of the states in Fig. 8.9c have been reversed. Here, the ordering between the rate constants would be $k(Ba^{2+}) < k(Ca^{2+}) > k(Mg^{2+})$ and the enzyme could thus be said to be optimized for the intermediate ion.

Calculations of the actual dependence of the activation barrier, $\Delta g^{\ddagger}$, on the metal size in the active site of SNase are summarized in Fig. 8.10. The results reflect mainly the energetics of ψ_2 and ψ_3, since the dependence on the ionic radius in ψ_1 is found to be rather small.

The origin of the dependencies of $\Delta \Delta g^{\ddagger}$ on r_{ion} can be rationalized in the following way. When the smaller metals are bound to the enzyme, the free energy of ψ_2 will be lowered considerably more than that of the transition state (as well as ψ_3) since in the former the OH^- ion is free to interact with or ligate the metal, while it is becoming partially bound to the 5'–P atom at the transition state with accompanying charge delocalization. On the other

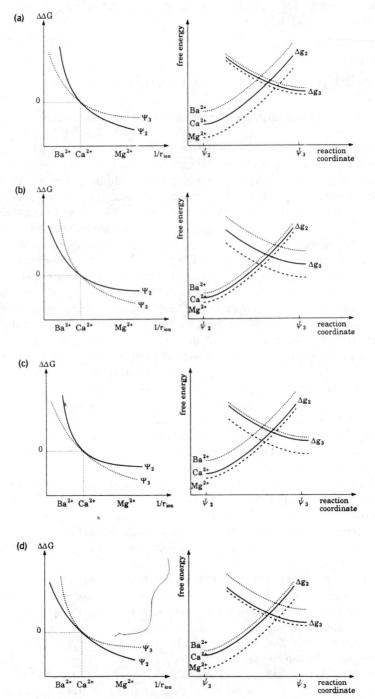

FIGURE 8.9. Linear free-energy relationship for the effect of metal substitution on ε_2 and ε_3 in staphylococcal nuclease (see text for details).

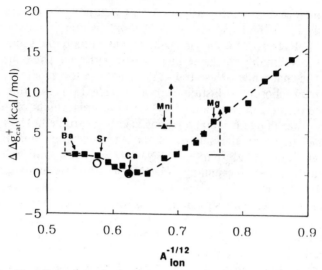

FIGURE 8.10. The effect of metal substitution on $\Delta\Delta g_{cat}^{\neq}$ on the catalytic reaction of SNase. The observed values of Sr^{2+} and Ca^{2+} are denoted by circles and the experimentally estimated limits for Ba^{2+}, Mn^{2+} and Mg^{2+} by ↑ (see Ref. 15 for more details).

hand, when the metal ion becomes too large it has less ability to perform its other major catalytic role (besides stabilizing the hydroxide ion in the first reaction step), namely, solvating the developing double negative charge on the phosphate group. That is, for the larger ions the state ψ_3 would be more sensitive to the ion size than ψ_2 because of the less efficient solvation of the phosphate group.

By calculating the quantities $\Delta\Delta G_1(Ca^{2+} \rightarrow M^{2+})$, $\Delta\Delta G_2(Ca^{2+} \rightarrow M^{2+})$, and $\Delta\Delta g_{2\rightarrow3}^{\ddagger}(Ca^{2+} \rightarrow M^{2+})$ it is possible to obtain the overall change in activation energy (relative to Ca^{2+}) as a function of the ion (M^{2+}) size. Such a calculation is presented in Fig. 8.10, where the location of Sr^{2+}, Ba^{2+}, Ca^{2+}, and Mg^{2+} have been indicated on the curve. The two main conclusions to be drawn from the dependence of $\Delta\Delta g_{cat}^{\neq}$ on the ion radius. First, that there is a clear minimum in the neighborhood of Ca^{2+}, which suggests that the enzyme has *been optimized to work exactly with calcium bound.* Secondly, it can be noted that the calculated effect on the catalytic rate is more pronounced when smaller ions, such as Mg^{2+}, replace Ca^{2+} than is the case for the larger Sr^{2+} and Ba^{2+} ion. This is mainly due to the fact that for smaller ions $\Delta\Delta G_2$ depends much more on the ion size than the corresponding free energies of the two other states, while for larger ions the free energy of all three states shows a more commensurable behavior. This trend appears to agree with the relevant experimental observations.

The finding that SNase appears to have its turnover optimum for the ion which it uses in nature may, of course, not be considered terribly surprising. However, the free-energy relationships leading to a rate optimization are quite interesting and point toward more general features pertaining also to other metalloenzymes, both with similar as well as quite different catalytic reactions. Perhaps the most immediate example is that of deoxyribonuclease I (DNase I) (Ref. 9). This enzyme catalyzes essentially the same reaction as SNase with presumably the same mechanistic pathway. The main difference appears to be that while SNase uses a glutamate as the general base, DNase I has instead chosen a histidine residue (His 131) for this step. The dependence of the catalytic rate of DNase I on replacement of the Ca^{2+} ion by various other divalent metal ions has also been studied. The influence of these replacements on the activity of the enzyme agrees qualitatively well with the calculated $\Delta\Delta g^{\ddagger}_{cat}$ curve for SNase (Fig. 8.10). Only Sr^{2+} and Ba^{2+} can replace the catalytic calcium ion in DNase I, but are less effective (Ba^{2+} more so than Sr).

Another example with similar mechanistic features, but for a different reaction, is the catalysis of ester bond hydrolysis in phosphoglycerides by phospholipase A_2. As for SNase and DNase I, phospholipase (Ref. 10) also has an absolute requirement for Ca^{2+} as a cofactor, and the Ca^{2+} appears to play a very similar role to that in SNase. It binds the negatively charged substrate phosphate group and probably also facilitates the abstraction of a proton to yield the OH^- nucleophile. Furthermore, it must be important for stabilizing the charges of the tetrahedrally coordinated C2 carbon transition state, in analogy with its multiple tasks in SNase. The proposed mechanism for phospholipase A_2 also involves general base-assisted catalysis in the first step of the reaction through an Asp–His pair similar to that found in the serine proteases (as well as DNase I). Several divalent metal ions have been shown to be inhibitory and no cation has been found that can replace Ca^{2+} in the enzymatic reaction. Since both Sr^{2+} and Ba^{2+} form ternary enzyme–metal–substrate complexes with phospholipase A_2, but neither ion promotes catalysis, it was suggested that only Ca^{2+} can effectively enhance polarization of the ester carbonyl oxygen in the second reaction step (as will be discussed at the end of this chapter, it is important to replace the somewhat useless concept of ground state bond polarization by the consideration of the electrostatic stabilization of the transition state). Thus, the reduced ability (compared to Ca^{2+}) for these larger ions to "solvate" the negatively charged transition state appears to provide a rationalization of the data also for phospholipase A_2, in manner similar to SNase (a less efficient stabilization of the OH^- nucleophile could also contribute to the absence of activity for these ions). However, the argument above cannot account for why the more electrophilic ions do not promote catalysis. For these ions, the inability to activate the enzyme may again reflect a strong interaction between the metal and the nucleophile, which hampers its possibility to attack the substrate.

Similar reaction mechanisms, involving general base and metal ion catalysis, in conjunction with an OH^- nucleophilic attack, have been proposed for thermolysin (Ref. 12) and carboxypeptidase A (Refs. 12 and 13). Both these enzymes use Zn^{2+} as their catalytic metal and they also have additional positively charged active site residues (His 231 in thermolysin and

Arg 127 in carboxypeptidase) with, presumably, similar transition state stabilization effects as the arginines in SNase, DNase I, and alkaline phosphatase. It is noteworthy that thermolysin and carboxypeptidase, as opposed to the previous cases, combine the choice of the Zn^{2+} ion, which increases the acidity of the reactive water molecule, with general base catalysis (by a glutamate), if the proposed mechanisms for these enzymes are correct. Metal substitution experiments on carboxypeptidase A have shown that the activity is optimal with Zn^{2+} or Co^{2+} bound. In this case the alkaline earth metals produce no activity. Interestingly, it appears that carboxypeptidase A is more sensitive to replacement of the Zn^{2+} ion by transition metals with larger hydration energy than by those with smaller hydration energy. This might be indicative of a free-energy relationship similar to that of Fig. 8.10, underlying the observed optimum for Co^{2+} and Zn^{2+}.

As a final example, consider the mechanistic features of the alcohol dehydrogenase (ADH)-catalyzed reaction (Ref. 14). This reaction differs somewhat from the previous cases, since the step following the alcohol deprotonation involves a hydride transfer rather than an $R-O^-$ nucleophilic attack. However, the deprotonation of the alcohol group corresponds to basically the same energetics *in solution* as the first step of the previous cases. That is, the free-energy cost of transferring the proton to water *in solution* is about 22 kcal/mol, and the enzyme must be able to reduce this energy to a much more tractable number in order to accomplish any catalysis at all. In this respect, it again appears that the Zn^{2+} ion bears the heaviest burden in catalyzing the first step of the reaction.

In all of the cases discussed above, the metal ion plays a central role in facilitating an otherwise unfavorable proton transfer step as well as in the subsequent transition-state stabilization and substrate binding. As for the first point above, it should be kept in mind that even with a general base (as opposed to a water molecule) to accept a proton from a water molecule, the cost of forming an OH^- nucleophile is about 11–16 kcal/mol *in solution*, depending on the type of general base (it is about 22 kcal/mol without general base catalysis). Therefore, the advantage of using a divalent metal ion in order to accelerate the first reaction step is obvious.

8.3.2. Classification of Metalloenzymes in Terms of the Interplay Between the General Base and the Metal

On the basis of the examples given above, it is reasonable to suggest that the underlying principles for optimization of the overall reaction rate with respect to the choice of metal ion are similar. That is, there are basically three states along the reaction pathway which determine the most suitable choice of metal ion. These are: (1) the reactant state with bound metal and substrate before the proton transfer step, (2) the intermediately created free OH^- nucleophile and, (3) the subsequent transition state associated with the nucleophilic attack. It must clearly be advantageous to reduce the cost of abstracting the proton from the nucleophile as much as possible, but, as elucidated in the case of SNase, a too electrophilic metal is likely to be less efficient by "trapping" the OH^- ion as a ligand. The electrostatic stabilization of the negatively charged transition state is not, at least in the case of SNase, as much affected by choosing a small electrophilic ion with large hydration energy as is the interaction with the free hydroxide ion. This is due to the higher degree of charge delocalization at the transition state, where the negative charge carried by the nucleophile is becoming distributed over several atoms.

It may be instructive to again consider the energetics of a proton transfer reaction of the type involved in the first step of the examples above, in solution. Under the influence of a possible general base as the proton acceptor and a possible metal ion assisting as a catalyst we can write

$$R - OH + B \overset{M}{\rightleftharpoons} R - O^- + BH^+ \tag{8.10}$$

where B is a base which can be either a water molecule or a stronger base, while M denotes a metal ion, if present, otherwise simply a water molecule. The energetics of eq. (8.10) (in solution) can be described by Fig. 8.11a, which shows the influence of some prototypes B and M on the reaction-free energy. The approximate numerical values in Fig. 8.11a are calculated from

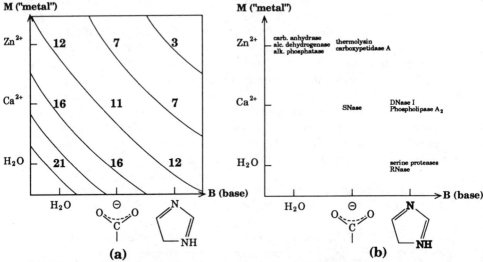

FIGURE 8.11. Classifying metalloenzymes according to their catalytic metal and the coupled general base. Part (a) of the figure shows the energetics (in kcal/mol) of transferring a proton from a metal-bound water to a general base in water. For example, a proton transfer from Ca^{2+}-bound water to glutamate costs 11 kcal/mol in water. Part (b) classifies different metalloenzymes according to the corresponding metal and general base. The figure illustrates that metalloenzymes are usually found in the low-energy part of the diagram.

observed pK_a-shifts in solution. If we think of Fig. 8.11a as defining a sort of free-energy surface for the solution reaction, it is interesting to examine to what extent this picture is reflected by enzymatic reactions of the same type. In Fig. 8.11b a number of enzymes with well-characterized reaction mechanisms are "plotted" according to their metal and general base. Although it is clear that the actual free-energy values of Fig. 8.11a cannot apply strictly to Fig. 8.11b (e.g., because of different dielectric properties in different active sites), it is probably significant that the "high-energy" region appears to be avoided in Fig. 8.11b.

Finally, it may be useful to comment here on the commonly used concept that relates the catalytic power of metal ions to their ability to *"polarize"* the reacting bond (e.g., the ester carbonyl in the reaction of phospholipase A_2). The concept of bond polarization is somewhat useless since it does not render itself to quantitative predictions. What really counts is the electrostatic interaction between the metal ion and the reacting fragments in their ground and transition state (e.g., $O^- C{=}O \cdots Ca^{2+}$ and $O{-}C{-}O^- \cdots Ca^{2+}$ in the phospholipase A_2 case). Once we define our mechanism in terms of the *energetics* of the fragments, rather than the ill-defined polarization concept, we can conveniently ask how much the given resonance form is stabilized and use linear free energy relationships in a semiquantitative way.

REFERENCES

1. F. A. Cotton, E. E. Hazen, and M. J. Legg, *Proc. Natl. Acad. Sci. U.S.A.*, **76**, 2551 (1979).

2. J. P. Guthrie, *J. Am. Chem. Soc.*, **99**, 3991 (1977).

3. E. H. Serpersu, D. Shortle, and A. S. Mildvan, *Biochemistry*, **25**, 68 (1986).

4. J. Aqvist and A. Warshel, *Biochemistry*, **28**, 4680 (1989).

5. D. N. Silverman and S. Lindskog, *Acc. Chem. Res.*, **21**, 30 (1988).

6. E. Magid and B. O. Turbeck, *Biochem. Biophys. Acta.*, **165**, 515 (1968).

7. A. E. Eriksson, P. M. Kylsten, T. A. Jones, and A. Liljas, *Proteins*, **4**, 283 (1988).

8. G. Eisenman and R. Horn, *J. Membr. Biol.*, **76**, 197 (1983).

9. (a) D. Suck and C. Oefner, *Nature (London)*, **321**, 620 (1986). (b) P. A. Price, *J. Biol. Chem.*, **250**, 1981 (1975).

10. H. M. Verheij, J. J. Volwerk, E. H. J. M. Jansen, W. C. Puyk, B. W. Dijkstra, J. Drenth, and G. H. de Haas, *Biochemistry*, **19**, 743 (1980). (b) B. W. Dijkstra, J. Drenth, and K. H. Kalk, *Nature (London)*, 604 (1981).

11. M. A. Wells, *Biochemistry*, **11**, 1030 (1972).

12. B. W. Matthews, *Acc. Chem. Res.*, **21**, 333 (1988).

13. D. W. Christianson, P. R. David, and W. N. Lipscomb, *Proc. Natl. Acad. Sci. U.S.A.*, **84**, 1512 (1987).

14. B. L. Vallee, A. Galdes, D. S. Auld, and J. F. Riordan, in *Zinc Enzymes*, T. G. Spiro (Ed.), Wiley, New York , 1983 p. 25.

15. J. Aqvist and A. Warshel, *J. Am. Chem. Soc.*, **112**, 2860 (1990).

16. J. Aqvist and A. Warshel (in preparation).

9

HOW DO ENZYMES REALLY WORK?

9.1. INTRODUCTION

The previous chapters taught us how to ask questions about specific enzymatic reactions. In this chapter we will attempt to look for general trends in enzyme catalysis. In doing so we will examine various working hypotheses that attribute the catalytic power of enzymes to different factors. We will try to demonstrate that computer simulation approaches are extremely useful in such examinations, as they offer a way to dissect the total catalytic effect into its individual contributions.

In searching for major catalytic effects one may start from Pauling's statement (Ref. 1) that enzymes catalyze their reactions by stabilizing the corresponding transition states. This statement reflects an early recognition that the transition state theory is applicable to enzymes and that the rate constant depends mainly on the activation free energy. This statement also led to the important prediction that transition state analogues would be good inhibitors. However, this early insight *does not* solve our problem. That is, it is very probable that most enzymes stabilize their transition states relative to the reference reaction in water, but the question is how this stabilization is accomplished. Many proposals have been put forward to rationalize the enormous catalytic power of enzymes (Refs. 2–11). In the following sections we will consider the main options.

9.2. FACTORS THAT ARE NOT SO EFFECTIVE IN ENZYME CATALYSIS

9.2.1. It Is Hard to Reduce Activation Free Energies in Enzymes by Steric Strain

The strain hypothesis, which was mentioned and discussed in Chapter 6, suggests that the steric force of the enzyme-active site reduces the activation-free energy by destabilizing the ground state. To estimate the actual magnitude of this effect we have to agree first on a common definition of "strain." Here we adopt the usual definition in *conformational analysis* and consider as steric potentials the repulsive van der Waals interactions and the contributions of bonds, bond angles, and torsional deformations. The charge–charge and charge-induced dipoles interactions are classified as electrostatic contributions, while the attractive van der Waals terms (whose effect in the protein, relative to the same process in water, is negligible) can be classified as either steric or electrostatic contributions. The main point in this definition is a clear division between the effects associated with electrostatic forces (which vary slowly with distance) and the effects associated with steric forces (that change fast with small molecular deformations).

With this definition we can assess the actual catalytic contribution associated with steric effects by a straightforward "computer experiment." That is, we can calculate the steric contribution to the activation free energy, $\Delta g_{\text{steric}}^{\neq}$, in both the enzyme site and in water. The difference $\Delta \Delta g_{\text{steric}}^{\neq} = (\Delta g_{\text{steric}}^{\neq})^p - (\Delta g_{\text{steric}}^{\neq})^w$ is the contribution of strain to the change in catalytic free energy. This type of calculation has been performed for the catalytic reaction of lysozyme (Chapter 6) and has indicated that the strain effect is not a major catalytic factor, since the protein is quite flexible and can accommodate the structural changes of the substrate without a large increase in free energy. This seems to be a quite general observation since the elementary steps in most chemical reactions do not involve large displacements of the reacting atoms (note that these displacements should be evaluated in a way that minimizes the change in their Cartesian coordinates for the given change in internal coordinates). It is still possible that some special reactions, that involve Cartesian displacements of more than 1 Å, may be associated with significant steric effects on $\Delta g^{\neq}$. However, such ground-state destabilization effects cannot help in increasing k_{cat}/K_M, which is (as is clearly illustrated in Fig. 5.2) only affected by the difference between the energy of the transition state, $ES^{\neq}$, and the energy of the E + S state. Thus these effects are less likely to be used in the evolutional development of enzymes, which is evolved under the requirement of optimal k_{cat}/K_M.

In some cases one finds that steric effects lead to clear changes in activity. The most obvious examples are the cases where the enzyme or the substrate are modified so that the reacting part of the substrate *cannot* assume the proper orientation in the active site. For example, introducing a bulky residue in the active site of trypsin can prevent an optimal orientation of the oxyanion intermediate in the oxyanion-hole. This effect, however, is not an example of a steric contribution to catalysis but of the construction of a bad catalyst. Another related example is the modification of a proton acceptor group in an enzyme that will pull it further away from the proton donor; for example, the reaction of *triosephosphate isomerase* involves a proton transfer from the dihydroxyacetone phosphate substrate to Glu-165. Mutation of Glu-165 to Asp leads to a reduction of the rate constant by a factor of about 1000 (see Ref. 14). Such a change can *reduce* drastically the rate constant due to steric restriction (this situation is illustrated in Fig. 9.1). Here again we do not have an example of the role of strain in enzyme catalysis, but of the role of strain in destroying enzyme activity. Both reactions in good enzyme and solution reactions will occur through pathway α and not through β, and the real issue is how to catalyze reactions that occur through pathway α.

Since steric effects can change catalysis (e.g., the above mentioned trypsin case), one may still argue that such effects do influence the correlation between structure and function. However, this case is not so relevant to structure–function correlation since the steric effects establish new structure and the activity associated with this structure is the main subject of our

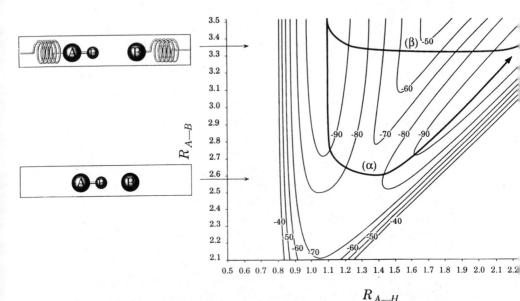

FIGURE 9.1. The potential surface for proton transfer reaction and the effect of constraining the R_{A-B} distance. The figure demonstrates that the barrier for proton transfer increases drastically if the $A - B$ distance is kept at a distance larger than 3.5 Å. However, in solution and *good* enzymes the transfer occurs through pathway α where the $A - B$ distance is around 2.7 Å.

discussion. Thus we conclude that while steric effects should clearly be considered and taken into account in correlating protein *sequence* and structure, they are not likely to provide a major catalytic advantage in most enzymes.

9.2.2. The Feasibility of the Desolvation Hypothesis Can Be Examined with Clear Thermodynamic Considerations

One of the interesting proposals for the origin of enzyme catalysis is the desolvation hypothesis (Ref. 7). According to this hypothesis, a nonpolar enzyme's active site can catalyze reactions by desolvating ground states which are strongly solvated in the corresponding reaction in solution. For example, in the S_N2 reaction of Fig. 9.2, a large part of the barrier is due to the loss of solvation energy associated with the formation of the delocalized charges of the transition state from the localized ground state charge. Moving the system to a nonpolar solvent will reduce the solvation energy of both the ground and the transition state by about half (see Exercise 9.1) and

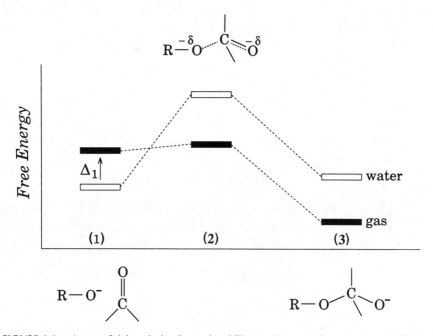

FIGURE 9.2. A superficial analysis of a nucleophilic attack on a carbonyl carbon in different environments. If a small value for the energy (Δ_1) associated with moving the charged R–O⁻ group from solution to the gas phase is assumed, one might conclude that the gas-phase reaction provides a reasonable model for the corresponding enzymatic reaction. However, a correct thermodynamic cycle will shift the gas phase reaction by $\Delta_1 = -\Delta g_{sol}$ (to a very high energy) and one will have to consider the barrier associated with the formation of R–O⁻ from R–OH (see Fig. 9.3 and Ref. 13).

the activation barrier will be reduced by about half. In fact, there are experimental demonstrations that some reactions can be accelerated by moving them from polar to nonpolar solvents (Refs. 5 and 7d, e). However, the analysis given above overlooks a major point; reactions in a nonpolar enzyme-active site are not the same as a reaction in a nonpolar solvent since the enzyme-active site is surrounded by a polar solvent. Thus the correct thermodynamic cycle for the reaction must include the energetics of forming the relevant fragments in aqueous solution and then moving them into the active site. This point is illustrated in Fig. 9.3 (see also Ref. 13). As is clear from the figure the apparent activation barrier includes the work of moving the charged O^- from water to the enzyme-active site and this amounts to a large (rather than small) barrier in a nonpolar enzyme.

Exercise 9.1. Evaluate the energetics of the reaction of Fig. 9.2 in a nonpolar enzyme-active site.

Solution 9.1. The energetics of this reaction in water is known from experimental information (Chapter 7). In order to estimate the corresponding energetics in a non polar site we start by expressing the electrostatic energy of a given state in a solvent of a dielectric constant d by (see Ref. 8a of Chapter 4).

$$\Delta g^i_{elec,d} = \Delta g^i_{sol,d} + V^i_{QQ} \simeq V^i_{QQ}/d + \Delta g^{i,\infty}_{sol,d} \qquad (9.1a)$$

where V_{QQ} is the electrostatic interaction between the reacting fragments in vacuum [see eq. (5.14)]. Next we use the Born's formula [eq. (3.21)] for the solvation energy of the fragments at infinite separation:

$$\Delta g^{i,\infty}_{sol,d} \simeq \sum_k \Delta G^{i,k}_{sol,w}\left[\left(1-\frac{1}{d}\right)\Big/\left(1-\frac{1}{80}\right)\right] \qquad (9.1b)$$

where $\Delta G^{i,k}_{sol,w}$ is the solvation energy of the kth fragment of the ith state in water. Using the above equations and neglecting terms which include the $1/d$ factor for the fragments in water, where $d \simeq 80$, we obtain

$$(\Delta\Delta g^i_{sol})_{w\to np} = (\Delta g^i_{sol,np} - \Delta g^i_{sol,w}) \simeq -\Delta g^i_{sol,w}/d_{np}$$

$$\simeq \left(V^i_{QQ} - \sum_k \Delta G^{i,k}_{sol,w}\right)\Big/d_{np} \qquad (9.1c)$$

where $\Delta g^i_{sol,w}$ and $\Delta g^i_{sol,np}$ are the solvation energies of the ith fragment in water and in a nonpolar site, respectively. With this we obtain

$$\Delta g^i_{np} = \Delta g^i_w + (\Delta\Delta g^i_{sol})_{w\to np} \simeq \Delta g^i_w + \left(V^i_{QQ} - \sum_k \Delta G^{i,k}_{sol,w}\right)\Big/2 \qquad (9.1d)$$

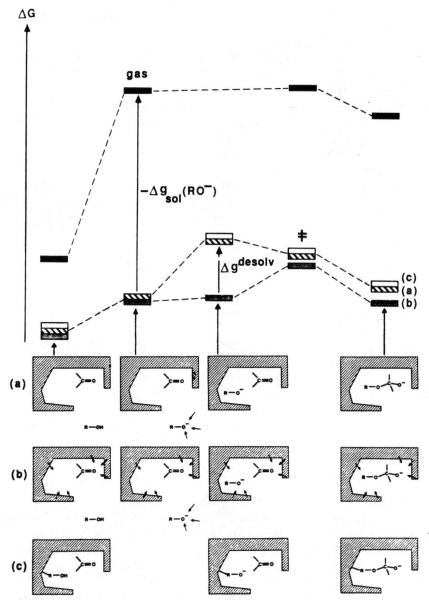

FIGURE 9.3. Illustrating why the desolvation mechanisms cannot lead to a lower activation barrier in enzymes, but possibly to a higher barrier. Three cases are compared: (*a*) formation of the charged nucleophile in water and its penetration to a nonpolar active site, (*b*) formation of the charged nucleophile in water and penetration to a polar active site, and (*c*) formation of the charged nucleophile in a nonpolar active site. The loss of solvation energy upon moving a R–OH group from water to a nonpolar active site is small compared to the corresponding change for a charged group. Therefore, the two cases (*a* and *c*) that correspond to a desolvation mechanism can both be described by the same diagram. The solvation substitution model (*b*), in which the charged groups are solvated effectively by the protein dipoles, will always give a lower activation barrier than a desolvation mechanism, since a desolvating active site inevitably will destabilize the R–O⁻ state more than the uncharged reference state and more than the charged state in solution.

where we use $d_{np} = 2$ and where Δg_w^i and Δg_{np}^i are the energies of the given state in water and in nonpolar sites, respectively. The solvation energies $\Delta G_{sol,w}^{i,k}$ can easily be obtained by the reader with the LD model and are frequently available from experimental studies (the values needed for the present problem are given in ref. 13). Using either the LD calculations or experimental estimates we obtain $(\Delta\Delta g_{sol}^{(1)})_{w\to np} \simeq 46 \text{ kcal/mol}$, $(\Delta\Delta g_{sol}^{(2)})_{w\to np} \simeq 32$. This gives $\Delta\Delta g_{w\to np}^{\neq} = (\Delta\Delta g_{sol}^{(2)})_{w\to np} - (\Delta\Delta g_{sol}^{(1)})_{w\to np} \simeq -14 \text{ kcal/mol}$, where the subscripts (1) and (2) are the corresponding states in Fig. 9.2.

This calculation demonstrates that a nonpolar solvent can accelerate S_N2 reactions. However, this is not what we are asking; the relevant quantity is the overall activation energy for the reaction in a nonpolar enzyme which is surrounded by water. Thus, as is indicated in the thermodynamic cycle of Fig. 9.3, we should include the energy of moving the ionized R–O$^-$ from water to the nonpolar active site $(\Delta\Delta g_{sol}^{(1)})_{w\to np}$. Thus the actual apparent change in activation barrier is

$$\Delta\Delta g^{\neq} \simeq (\Delta\Delta g_{sol}^{(1)})_{w\to np} + \Delta\Delta g_{w\to np}^{\neq} \simeq 46 - 14 \simeq 32 \text{ kcal/mol} \qquad (9.2)$$

The main point of this exercise and considerations is that you can easily examine the feasibility of the desolvation hypothesis by using well-defined thermodynamic cycles. The only nontrivial numbers are the solvation energies, which can however be estimated reliably by the LD model. Thus for example, if you like to examine whether or not an enzymatic reaction resembles the corresponding gas-phase reaction or the solution reaction you may use the relationship

$$\Delta g_{gas}^i = \Delta g_w^i - \Delta g_{sol,w}^i \qquad (9.3)$$

Using this relationship for different enzymatic reactions (e.g., Ref. 13) indicates that enzymes *do not* use the desolvation mechanism and that their reactions have no similarity to the corresponding gas-phase reaction, but rather to the reference reaction in water. In fact, enzymes have evolved as better solvents than water, by providing an improved solvation to the transition state (see Section 9.4).

One may still conceive cases where destabilization of charged ground states can contribute to catalysis, and where nonelectrostatic binding forces (e.g., hydrophobic forces) compensate for the energy of moving the charges to the enzyme-active site. However, most of the regular functional groups in proteins (e.g., ionizable amino acids) will become unionized when placed in nonpolar active sites. Thus, for example, with a neutral ground state we will have to pay for ionizing the relevant groups in a nonpolar environment (e.g., Fig. 9.3c). More importantly, enzymes that have evolved in order to optimize k_{cat}/K_M could not benefit from destabilizing ground states charges, but only from stabilizing the charges of the transition states (see Fig. 5.2).

Thus it is concluded that while destabilization of the ground-state charges may be used in enzymes to reduce $\Delta g^{\neq}$, it is not used in enzymes that optimize k_{cat}/K_M. Furthermore, we argue that the feasibility of any proposed desolvation mechanism can be easily analyzed (and in most cases disproved) by the reader once the relevant thermodynamic cycle is defined and the solvation energies of the reacting fragments are estimated.

9.2.3. Dynamical Effects and Catalysis

It has been frequently suggested that dynamical factors are important in enzyme catalysis (Ref. 9), implying that enzymes might accelerate reactions by utilizing special fluctuations which are not available for the corresponding reaction in solutions. This hypothesis, however, looks less appealing when one examines its feasibility by molecular simulations. That is, as demonstrated in Chapter 2, it is possible to express the rate constant as

$$k = \bar{\tau}^{-1} \exp\{-\Delta G^{\neq}\beta\} \tag{9.4}$$

where we use here the rigorous rate expression with $\Delta G^{\neq}$, rather than the approximate expression with the $\Delta g^{\neq}$ of eq. (3.31), since we would also like to explore entropic effects. The inverse time τ^{-1} is the only part of the rate constant that reflects dynamical effects, while the activation-free energy $\Delta G^{\neq}$ reflects the nondynamical thermodynamic probabilities. Thus the issue here is whether an enzyme can increase $\bar{\tau}^{-1}$ in a significant way.

This question can be explored within the *linear-response* approximation, which relates the response of the effective coordinate of the environment (e.g., the solvent or the protein) to the dipole, μ, of the solute by (Ref. 11)

$$\langle Q(t) \rangle = (Q_{max}/\mu_{max}) \int_0^t \frac{\langle \dot{Q}(0)Q(t') \rangle}{\langle \dot{Q}(0)Q(0) \rangle} \langle \mu(t-t') \rangle \, dt' \tag{9.5a}$$

$$\bar{\tau}^{-1} = (\partial \langle Q(t)_+ \rangle / \partial t)/\Delta Q^{\neq} \tag{9.5b}$$

where Q is the generalized solvent coordinate which is defined as the solvent contribution to the energy gap $\varepsilon_j(t) - \varepsilon_i(t)$ for a reaction which involves a transfer from the potential surface ε_i to ε_j. As explained in detail in Ref. 11, this coordinate is related to the projection of the field from the solvent on the solute dipole. Eq. (9.5) can be used to evaluate the average time dependence of the solvent coordinate in a reactive trajectory. In doing so it is useful to obtain the time dependance of the solute dipole from several downhill trajectories and to approximate the calculated autocorrelation function $\langle Q(0)Q(t) \rangle$ by a single exponential function:

$$\langle Q(0)Q(t) \rangle \simeq B \exp\{-t/\tau_Q\} \tag{9.6}$$

while using the relationship $-\langle \dot{Q}(0)Q(t) \rangle = \partial \langle Q(0)Q(t) \rangle / \partial t$. (The characteristic time τ_Q is frequently referred to as the *longitudinal dielectric relaxation time* of the solvent). In the frequent case where τ_Q is shorter than the relaxation time of the solute dipole one finds (Ref. 11) that τ_Q determines $\bar{\tau}$.

When the approximation of eq. (9.6) is not justified, or when the relaxation time of μ is slower than τ_Q, we may determine $\bar{\tau}^{-1}$ in a direct way by eq. (9.5b).

An examination of the autocorrelation function $\langle Q(0)Q(t) \rangle$ and the corresponding τ_Q for the nucleophilic attack step in the catalytic reaction of subtilisin is presented in Fig. 9.4. As seen from the figure, the relaxation times for the enzymatic reaction and the corresponding reference reaction in solution are not different in a fundamental way and the preexponential factor $\bar{\tau}^{-1}$ is between 10^{12} and 10^{13} sec^{-1} in both cases. As long as this is the case, it is hard to see how enzymes can use dynamical effects as a major catalytic factor.

The above arguments can be restated in terms of related formulations (e.g., Ref. 15, Ref. 16 and Appendix A of Ref. 11) that explore in a somewhat more formal way the role of dynamical effects in chemical

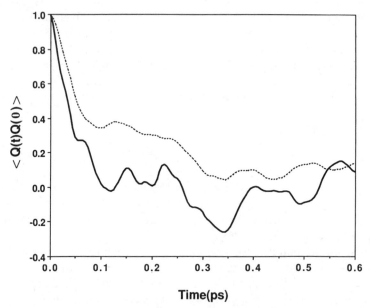

Time(ps)

FIGURE 9.4. The autocorrelation function of the time-dependent energy gap $Q(t) = (\varepsilon_3(t) - \varepsilon_2(t))$ for the nucleophilic attack step in the catalytic reaction of subtilisin (heavy line) and for the corresponding reference reaction in solution (dotted line). These autocorrelation functions contain the dynamic effects on the rate constant. The similarity of the curves indicates that dynamic effects are not responsible for the large observed change in rate constant. The autocorrelation times, τ_Q, obtained from this figure are 0.05 ps and 0.07 ps, respectively, for the reaction in subtilisin and in water.

reactions in solutions. These formulations predict rather small dynamical effects (factors of 10 in the most extreme cases, as long as one deals with reactions whose activation barriers are more than 5 kcal/mol), while we are interested in rate acceleration of many orders of magnitude. Furthermore, using the τ_Q's of Fig. 9.4 in the expressions of Refs. 15 and 16, one obtains negligible differences between the rate constants of reactions in enzymes and the corresponding reactions in solutions.

9.3. WHAT ABOUT ENTROPIC FACTORS?

It has been frequently proposed that enzymes catalyze reactions by using entropic effects (Refs. 3–5). This idea, which has been put forward in different ways, implies that the ground-state free energy is raised by fixing the reactants and products in an exact orientation and that this is a major catalytic effect. In exploring entropic effects one has to be quite careful in defining the problem correctly. In particular, the definition of the proper reference state is crucial. If, for example, we take our solvent cage as a reference state (Exercise 5.1), the concentration factors associated with bringing the reactants to the same cage are eliminated and one is left with true entropic factors which are the subject of this section.

In exploring the entropic difference between a given enzyme and its reference solvent cage, we should consider the dependence of the activation barrier on the activation entropy using the relationship

$$\Delta\Delta G^{\neq}_{s\to p} = \Delta G^{\neq}_{p} - \Delta G^{\neq}_{s} = \Delta\Delta H^{\neq}_{s\to p} - T\Delta\Delta S^{\neq}_{s\to p}$$

$$\Delta\Delta S^{\neq}_{s\to p} = \Delta S^{\neq}_{p} - \Delta S^{\neq}_{s} = (S^{\neq}_{p} - S^{0}_{p}) - (S^{\neq}_{s} - S^{0}_{s}) \qquad (9.7)$$

where S^0 designates the entropy in the reactant state.

As is obvious from Eq. (9.7), it is possible (at least in principle) to reduce $\Delta\Delta G^{\neq}$ by reducing S^0_p or by increasing $S^{\neq}_p$. Exploring whether such effects really occur in proteins is far from simple. A unique experimental demonstration that a given catalytic effect is associated with an entropic factor (e.g., the restriction of the ground-state configurations by the enzyme) is not available, and computer simulation approaches are not so effective at the present time (since the convergence of calculations of entropic contributions is still rather poor). Thus we will explore here the feasibility of entropic catalysis in a somewhat qualitative way, using sometimes simple logical arguments.

9.3.1. Entropic Factors Should be Related to Well-Defined Potential Surfaces

In order to explore the significance of entropic factors, we must relate the different hypotheses to the clear concept of potential surfaces. Thus we start

by taking the simple example of the nucleophilic attack reaction
($O^-C=O \rightarrow O-C-O^-$) in amide hydrolysis and demonstrate the relationship
between the reaction potential surface and the entropic contributions. The
approximated EVB potential surface for this reaction in solution is drawn in
Fig. 9.5, using equipotential lines (contours) with increments of 0.6 kcal/mol
(which corresponds to β^{-1} at room temperature). The activation free energy
for this surface can be estimated by

$$\exp\{-\Delta G^{\neq}\beta\} = z^{\neq}/z_0 = \int_{-\infty}^{\infty} \int_{X^{\neq}-\Delta X^{\neq}/2}^{X^{\neq}+\Delta X^{\neq}/2} e^{-U\beta} \, dX \, ds \Big/ \int_{-\infty}^{\infty} \int_{-\infty}^{X^{\neq}} e^{-U\beta} \, dX \, ds$$

$$\simeq \left(\sum_{i(R^{\neq})} e^{-U_i\beta} \, \Delta v_i\right) \Big/ \left(\sum_{j(R_0)} e^{-U_j\beta} \, \Delta v_j\right) \qquad (9.8)$$

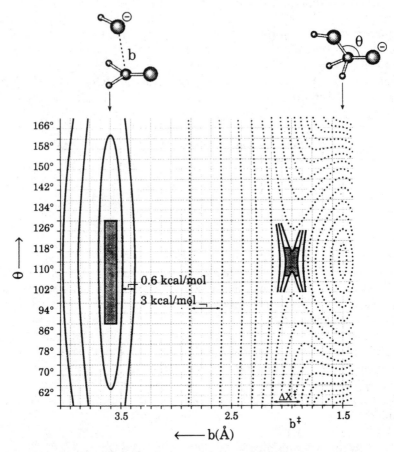

FIGURE 9.5. The potential surface for the $O^-C=O \rightarrow O-C-O^-$ step in amide hydrolysis in
solution, where the surface is given in terms of the angle θ and the distance b. The heavy
contour lines are spaced by β^{-1} (at room temperature) and can be used conveniently in
estimating entropic effects. The figure also shows the regions (cross hatched) where the
potential is less than β^{-1} for the corresponding reaction in the active site of subtilisin.

where s designates the coordinates perpendicular to the reaction coordinate X (θ and b are taken in the present case as s and X, respectively). Here $R^{\neq}$ designates the transition state region, R_0 designates the reactant region (as indicated by the limits of the corresponding integrals) and the Δv's are small volume elements in the given space. This equation gives $\Delta G^{\neq}$ in terms of the ratio between the partition functions of the transition state and the reactant states, which can be estimated easily by counting the available configurations with low potential energy in both states.

In the following section we will only consider the contribution to z_0 from the configurations which are within the solvent cage region (the remaining contributions are evaluated in Exercise 5.1). Thus we will be focusing on entropic contributions to $\Delta g_{cat}^{\neq}$ rather than $\Delta G^{\neq}$.

Exercise 9.2. Estimate $\Delta G^{\neq}$ and $\Delta S^{\neq}$ for the system in Fig. 9.5.

Solution 9.2. The $\Delta G^{\neq}$ of eq. (9.8) can be estimated by including in the relevant sum only those terms that are within $2\beta^{-1}$ from the lowest point in the corresponding term (higher-energy regions will give only small contributions). Thus we can simply count the squares with the given value of U and use the volume element ($\Delta\theta \sin\theta \, \Delta b$), replacing $\sin\theta$ by its value in the center of the corresponding square. This gives

$$z^{\neq} \simeq (4e^{-U^{\neq}\beta} + 6e^{(-U^{\neq}+\beta^{-1})\beta}) \Delta\theta \, \Delta b$$

$$\simeq e^{-U^{\neq}\beta}(4 + 6e^{-1}) \Delta\theta \, \Delta b \simeq 6\Delta\theta \, \Delta b e^{-U^{\neq}\beta}$$

$$z_0 \simeq (40e^{-0\beta} + 60e^{-\beta^{-1}\beta}) = (40 + 60e^{-1}) \Delta\theta \, \Delta b \simeq 60\Delta\theta \, \Delta b \quad (9.9)$$

The resulting $\Delta G^{\neq}$ is given by

$$\Delta G^{\neq} \simeq -\beta^{-1} \ln(z^{\neq}/z_0) \simeq U^{\neq} - \beta^{-1} \ln(6/60) \quad (9.10)$$

The second term in eq. (9.10) is the $T \Delta S^{\neq}$ term $[-T \Delta S^{\neq} = -\beta^{-1}\ln(6/60)]$ and we obtain

$$-T \Delta S^{\neq} \simeq -\beta^{-1} \ln(6/60) \quad (9.11)$$

Realizing that the main contribution to eq. (9.7) comes from terms within the β^{-1} counter line we may use the approximation

$$-T \Delta S^{\neq} \simeq -\beta^{-1} \ln(v^{\neq}/v^0) \quad (9.12)$$

where $v^{\neq}$ and v^0 designate the configurational volumes in $R^{\neq}$ and R^0, respectively, whose potential energies are β^{-1} or less above the lowest potential in the given range.

After learning to estimate $\Delta G^{\neq}$ and $\Delta S^{\neq}$, we might ask how $\Delta\Delta S_{s\to p}^{\neq}$ is affected by the steric restriction of the protein environment. As is clear from eq. (9.7), we need the *differences* between the entropic contributions to $\Delta G^{\neq}$ rather than the individual $\Delta S^{\neq}$. This requires the examination of the difference between the potential surfaces of the protein and solution reaction. Here we exploit the fact that the electrostatic potential changes rather slowly and use the approximation

$$U_p^{\neq} \simeq U_s^{\neq} + U_{p,\text{strain}}^{\neq} + \Delta^{\neq}$$

$$U_p^0 \simeq U_s^0 + U_{p,\text{strain}}^0 + \Delta^0 \qquad (9.13)$$

where the Δ's are the relatively constant contributions from electrostatic interactions to the difference between U_p and U_s. Here we assume that there are no significant steric forces in the solvent cage (the solvent should be allowed to relax for each solute configuration in proper calculations of $\Delta G^{\neq}$). In our specific example of the $O^-C=O \to O-C-O^-$ reaction in subtilisin, we find that $U_{p,\text{strain}}$ is less than β^{-1} at $\theta = 105 \pm 30°$ and is larger than β^{-1} outside this range (this steric potential is indicated in Fig. 9.5). With the above U_{strain} one finds that the available configuration space in the protein's transition state is not much different than the corresponding space in solution, but the ground-state configuration space v_p^0 and v_s^0 are different. This gives

$$- T \,\Delta\Delta S^{\neq} \simeq -\beta^{-1} \ln[(v_p^{\neq}/v_p^0)/(v_s^{\neq}/v_s^0)] \simeq -\beta^{-1} \ln(v_s^0/v_p^0) \quad (9.14)$$

In our specific example $(v_s^0/v_p^0) \simeq 40/24$ and $-T\,\Delta\Delta S^{\neq} \simeq -0.6\,\text{kcal/mol}$.

With this insight in mind you might examine the so-called *orbital steering mechanism* (Ref. 3). This interesting hypothesis considers the possibility that the transition state energy is a very steep function of the overlap between the orbitals of the reacting fragments (very small $v_s^{\neq}$ in our notation). The overall proposal has not been rigorously formulated, in both the original work and subsequent discussions by other workers, in terms of the well-defined parameters $v_p^{\neq}$, $v_s^{\neq}$, v_p^0, and v_s^0, but it has been implied that the enzyme keeps the reacting fragments in the exact orientation for the optimal transition state. This means, in terms of our more accurate concepts, that $v_p^0 \simeq v_s^{\neq}$. Thus it is implicitly assumed that $-T\,\Delta\Delta S^{\neq} \simeq -\beta^{-1}\ln(v_s^0/v_s^{\neq})$. Assuming that $v_s^{\neq}$ is very small gives very large entropic factors through this expression. The validity of the assumption is examined in the following exercise.

Exercise 9.3. Determine the entropic contributions to $\Delta\Delta G^{\neq}$ in the orbital steering model, using (a) $v_s^{\neq} = \Delta b \times 0.1^0$, $v_s^0 \simeq \Delta b \times 40°$ and (b) the EVB estimate of $v_s^{\neq}$ for the $O^-C=O \to O-C-O^-$ reaction. Note that this model implies that $v_p^0 \simeq v_s^{\neq}$.

Solution 9.3. (a) With the estimate $v_s^0/v_s^{\neq} = 40/0.1$ we obtain $-T\,\Delta\Delta S^{\neq} \simeq -\beta^{-1}\ln(v_s^0/v_s^{\neq}) = -0.6\ln(40/0.1) = -3.6\,\text{kcal/mol}$ which is a very large factor. (b) This result should, however, be reexamined with a realistic (rather than hypothetical) estimate of $v_s^{\neq}$. This can be done by the EVB formulation, noting that the transition-state potential is given by $U^{\neq} \simeq \frac{1}{2}(\varepsilon^{(2)} + \varepsilon^{(1)}) - H_{12}$ where $\varepsilon^{(1)}$ and $\varepsilon^{(2)}$ are, respectively, the potentials for the $O^-C=0$ and $O-C-O^-$ resonance structures. Since the θ dependance of $\varepsilon^{(1)}$ and H_{12} is small (there is no bond between O and C in this configuration), we can write $\Delta U^{\neq}(\theta) \sim \frac{1}{2}\Delta\varepsilon^{(2)}(\theta) \simeq 1.6 \cdot 10^{-2}(\theta - \theta_0)^2 \,\text{kcal mol}^{-1}\,\text{degree}^{-2}$ where we took a typical X–C–X bending force constant from Table 4.2 and converted it to the current units. Now we can determine $v_s^{\neq}$ by requiring $\Delta U^{\neq}$ to be equal to β^{-1} or $0.6\,\text{kcal/mol}$. This can be written as $1.6 \cdot 10^{-2}\Delta\theta^2 = 0.6$, which gives $v_s^{\neq} \simeq \Delta b \times 6^0$, and much smaller entropic contributions than for $v_s^{\neq} \simeq \Delta b \times 0.1^0$.

As is clear from this discussion and exercise, one can estimate $v_s^{\neq}$ in a realistic way. However, the correct estimate of $\Delta\Delta S^{\neq}$ requires a clear definition of the problem considering the available configurations $v^{\neq}$ and v^0 in both protein and solution. For example, it appears that v_p^0 is much *larger* than what was assumed in early works, since proteins are quite flexible. Thus, even if $v_s^{\neq}$ is very small, it does not mean that $\Delta\Delta S^{\neq}$ is large, since the assumption of $v_p^0 \sim v_s^{\neq}$ is invalid. It is interesting to note that with an unrealistically rigid protein, where v_p^0 is much smaller than $v_s^{\neq}$, we will find that the same steric effect of the protein will also make $v_p^{\neq}$ very small (with shallow $v_s^{\neq}$ we will find that $v_p^{\neq}$ is determined by the protein strain and is given approximately by v_p^0). This will give $(v_p^{\neq}/v_p^0) \sim 1$ and $v_s^{\neq}/v_s^0$ will determine $\Delta\Delta S^{\neq}$.

This discussion demonstrates the need for a clear definition of different entropic hypotheses in terms of well-defined potential surfaces which can then be examined by clear thermodynamic concepts.

9.3.2. Entropic Factors in Model Compounds and Their Relevance to Enzyme Catalysis

The entropic hypothesis seems at first sight to gain strong support from experiments with model compounds of the type listed in Table 9.1. These compounds show a huge rate acceleration when the number of degrees of freedom (i.e., rotation around different bonds) is restricted. Such model compounds have been used repeatedly in attempts to estimate entropic effects in enzyme catalysis. Unfortunately, the information from the available model compounds is not directly transferable to the relevant enzymatic reaction since the observed changes in rate constant reflect interrelated factors (e.g., strain and entropy), which cannot be separated in a unique way by simple experiments. Apparently, model compounds do provide very useful means for verification and calibration of reaction-potential surfaces

TABLE 9.1. Relative Rates for the Ring-Closure Reactions ($R'COO^-$ +
$R^HCOOR \rightarrow R'COOCOR'' + {}^-OR$) of Related Model Compounds, Which Can Be Used in
Estimating the Importance of Entropic Effects in Solution Reactions (see Ref. 2)

n	Compound	k_{rel}
1	$CH_3COO^{\ominus}+CH_3COOC_6H_5Br$	1.0
2	$COOC_6H_4Br$ / $COO^{\ominus}$	~$1 \times 10^3 M$
3	$COOC_6H_4Br$ / $COO^{\ominus}$	$3 \times 10^3 M$ — $1.3 \times 10^6 M$
4	$COOC_6H_4Br$ / $COO^{\ominus}$	~$2.2 \times 10^5 M$
5	$COOC_6H_4Br$ / $COO^{\ominus}$	$1 \times 10^7 M$
6	$COOC_6H_4Br$ / $COO^{\ominus}$	~$5 \times 10^7 M$

and simulation methods, but they cannot be used in a direct assessment of
entropic factors in enzymatic reactions. In other words, the potential
surfaces and the simulations probably provide the best way of analyzing and
transferring the information from model compounds to enzymes. With this
in mind, we will consider here only one simple example of the information
from intramolecular reactions in model compounds by examining the difference between compounds (4) and (5) in Table 9.1. The dependence of the
potential surface of these molecules on the central dihedral angle (ϕ_2) and
the $O^- \cdots C$ distance (b) is estimated in Fig. 9.6. The value of the potential
surface for each ϕ_2 and b was determined by minimizing the energy of the
system with respect to all other coordinates. As in Exercise 9.2 one can use
the resulting surface and eq. (9.8) to estimate the relevant entropic effect by
counting the volume elements with $U \le \beta^{-1}$ in the reactants and transition-
state regions. This gives $v^{\neq}/v_0 \simeq 10/500$ and $1/4$ for compounds (4) and (5)
respectively. Thus we obtain $-T \Delta\Delta S^{\neq}_{4\rightarrow5} = -\beta^{-1} \ln(50/4) \simeq -1.4$ kcal/

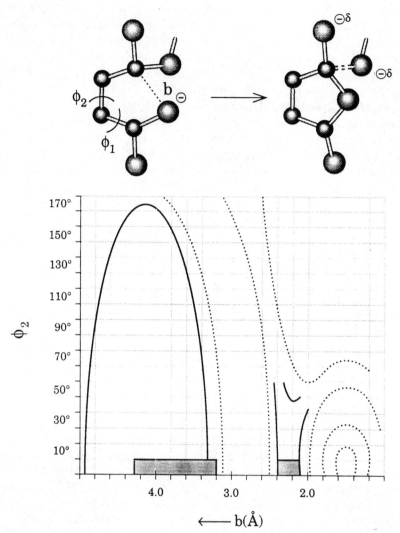

FIGURE 9.6. Analyzing the contributions from entropic effects to the free energy of ring closure reactions. The actual reaction involves a displacement of a $^-$OR group, but we only consider here the formation of the transition state. The figure displays the potential surface of compound (4) in terms of ϕ_2 and b, where the energy of the system is minimized at each point with respect to all other coordinates. The heavy and the dotted contour lines are spaced by β^{-1} and $10\beta^{-1}$, respectively (at room temperature), and can be used in estimating entropic effects. The figure also gives the regions (shaded) where the potential is less than β^{-1} for the corresponding reaction in compound (5). In this case the rotation around ϕ_2 is drastically restricted.

mol. While the corresponding observed ratio between the rate constants gives $\Delta\Delta G^{\neq}_{obs} = -\beta^{-1} \ln 50 = -2.3 \, kcal/mol$. A better agreement is obtained by a more rigorous treatment that counts all the available configurations with $\Delta U \leq \beta^{-1}$, including those associated with ϕ_1 and ϕ_3. Such a treatment (that cannot be displayed in a simple two-dimensional potential surface) can be easily performed. One can also use free-energy perturbation approaches for estimating the relevant $\Delta\Delta G^{\neq}$.

The above discussion demonstrates that significant entropic effects do indeed operate in ring closure reactions. This fact might imply that enzymes produce enormous entropic effects by fixing the reacting fragments (that might be viewed as the analogues of the ends of the chains involved in our ring closure reactions). This, however, is *not* directly related to regular enzymatic reactions since many configurations that are being restricted upon ring closure would not be so relevant to the difference between enzymatic reactions and the corresponding intermolecular reactions. For example, a large fraction of the additional configuration space of compound (4) [relative to compound (5)] occurs with large values of b that will place the corresponding intermolecular reaction out of our reference solvent cage (the contribution of these configurations to $\Delta G^{\neq}$ is already considered in our concentration calculations). In fact, the considerations of Fig. 9.5 are more relevant to the difference between the intermolecular reaction and the corresponding enzymatic reaction than those of Fig. 9.6. Apparently we do not have, as yet, direct experimental information about the magnitude of the entropic contribution to enzyme catalysis (which might indeed be significant). This emphasizes the need for computer simulations in assessing the importance of the rather complicated entropic factors.

It might be important to comment here on the hypothesis of Page and Jencks (Ref. 4) that received significant attention in the literature. This hypothesis implies that enzyme catalysis is due to the loss of rotational and translational entropy of the reacting fragments upon transfer from solution to the enzyme-active site. However, although this could be a significant factor in catalysis, it is probably overestimated. That is, Page and Jencks estimate the entropic contribution as that associated with the complete loss of rotational and translational degrees of freedom of the reacting fragments. However, the rotational and translational degrees of freedom are converted in the enzyme active site to low frequency vibrational mods with significant entropic contributions. It is clear now that the enzyme substrate complex is not as rigid as previously thought and no degree of freedom is completely frozen. This is why we formulated the problem in terms of the available volumes $v^{\neq}$ and v_0. Evaluating these volumes or related simulation approaches, should allow one to really examine what is the actual entropic contribution (in addition to the trivial cage effect estimated in Exercise 5.1). Reformulating the Page and Jencks hypothesis in terms of the more precise approach of eq. (9.14) one finds that the relevant $\Delta\Delta S^{\neq}$ should only include those degrees of freedom whose available space is drastically reduced at the transition state. For others, such as the rotation around the bond b in Fig.

9.5, one finds similar steric restrictions at the ground and transition state in the enzyme-active site. The corresponding contribution to $\Delta\Delta S^{\neq}$ is small. Furthermore, while fixing the reacting fragments might change the $\Delta g^{\neq}$ that corresponds to k_{cat}, it is hard to see how such an effect can change the $\Delta G^{\neq}$ that corresponds to k_{cat}/K_M. In fact, fixing the reacting fragments *decreases* the entropy of the transition state (this effect is not significant if the reacting fragments are also fixed at the transition state of the reference solvent cage).

In summary, as shown above, the discussion of entropic factors might be very complicated and involves major semantic problems, such as the definition of the relevant reference state. Thus it is essential to be able to *calculate* the actual entropic contribution to $\Delta G^{\neq}$ with well-defined potential surfaces. At present it does not seem likely that converging calculations of $\Delta\Delta S^{\neq}$ will attribute very large catalytic effects to true entropic factors, but more studies are clearly needed. It should be noted, however, that calculations of entropic effects in active sites of enzymes may be simpler than calculations of such effects in model compounds. This is why we chose as a reference state a solvent cage where the reacting fragments are in the same general orientation as in the enzyme. This procedure can be viewed as a practical way of using experimental information about the reacting fragments to extract the different gas phase parameters (the α_i's, and the H_{ij}'s), while avoiding the need to calculate ΔS_s^0 and to study the real solution reactions. With reliable α_i's, we can calculate the $\Delta g_p^{\neq}$ for our enzymatic reaction without facing the challenge of calculating entropic effects in the solution reaction. The entropic contributions to $\Delta g_p^{\neq}$ may be estimated by the FEP approach, provided that fragments are confined to several well defined regions. However, a more systematic study of entropic effects in both the enzyme and the solvent cage should involve considerations of the available low energy configurations (see Section 9.3.1).

9.4. ELECTROSTATIC ENERGY IS THE KEY CATALYTIC FACTOR IN ENZYMES

9.4.1. Why Electrostatic Interactions Are So Effective in Changing $\Delta\Delta g^{\neq}$

As discussed and demonstrated in the previous chapters, the catalytic effect of several classes of enzymes can be attributed to electrostatic stabilization of the transition state by the surrounding active site. Apparently, enzymes can create microenvironments which complement by their electrostatic potential the *changes* in charges during the corresponding reactions. This provides a simple and effective way of reducing the activation energies in enzymatic reactions.

In order to examine what makes electrostatic stabilization more effective than other feasible factors, it is useful to ask what is required for an effective reduction of $\Delta\Delta g_{s\rightarrow p}^{\neq}$. We may start from the general statement that an effective catalyst must interact with the *changes* during the reaction and such changes can be classified according to the following three classes: (1) changes in structures, (2) changes in available configurations, and (3) changes in charges. The structural changes in the elementary steps of most chemical reactions are relatively small and, as discussed before, cannot lead to large steric contributions to $\Delta\Delta g^{\neq}$ (since the steric potentials are steep and can be relaxed by small displacements of the protein atoms). The changes in the available configurations and the corresponding entropic contributions are also ineffective in reducing $\Delta\Delta g^{\neq}$ (see Section 9.3). On the other hand, the changes in charge distribution during the reaction can be translated to significant changes in $\Delta\Delta g^{\neq}$, since the electrostatic potentials are not very steep and can be used to "store" large energy contributions.

As discussed in the early sections it seems that there are very few effective ways to stabilize the transition state and electrostatic energy appears to be the most effective one. In fact, it is quite likely that any enzymatic reaction which is characterized by a significant rate acceleration (a large $\Delta\Delta g^{\neq}_{s \to p}$) will involve a complimentarity between the electrostatic potential of the enzyme-active site and the change in charges during the reaction (Ref. 10). This point may be examined by the reader in any system he likes to study.

The concept of electrostatic complimentarity is somewhat meaningless without the ability to estimate its contribution to $\Delta\Delta g^{\neq}$. Thus, it is quite significant that the electrostatic contribution to $\Delta\Delta g^{\neq}$ that should be evaluated by rigorous FEP methods can be estimated with a given enzyme–substrate structure by rather simple electrostatic models (e.g., the PDLD model). It is also significant that calculated electrostatic contributions to $\Delta\Delta g^{\neq}$ seem to account for its observed value (at least for the enzymes studied in this book). This indicates that simple calculations of electrostatic free energy can provide the correlation between structure and catalytic activity (Ref. 10).

9.4.2. The Storage of Catalytic Energy and Protein Folding

The previous section suggested that the catalytic power of enzymes is related to their ability to stabilize the changes in the reactant charges during the reaction. It might be argued, however, that the same stabilization effect can be obtained in other polar solvents (e.g., water) that can reorient their dipoles toward the transition-state charge distribution. For example, the interaction potential between the oxyanion transition state of amide hydrolysis and its surrounding solvent cage is not much different than the corresponding interaction with the oxyanion-hole in trypsin. The two cases, however, are quite different. In the enzyme the stabilizing dipoles are *preoriented* in the ground state toward the transition-state charges. In solution, on the other hand, it costs significant energy to orient the solvent dipoles to their transition-state configuration. In general, one finds that about half of the free energy associated with the charge–dipole interactions, $\Delta G_{Q\mu}$, is spent on the dipole–dipole repulsion, $\Delta G_{\mu\mu}$, so that

$$\Delta G_{\text{sol}} = \Delta G_{Q\mu} + \Delta G_{\mu\mu} \simeq \frac{1}{2}\, \Delta G_{Q\mu} \qquad (9.15)$$

In proteins, however, a significant part of $\Delta G_{\mu\mu}$ (or the corresponding reorganization energy of Chapter 3) is already paid for in the folding process, where the folding energy is used to compensate for the dipole–dipole repulsion energy and to align the active-site dipoles in a way that will maximize $\Delta G_{Q\mu}$. With preoriented dipoles we do not have to pay a significant part of $\Delta G_{\mu\mu}$ during the formation of the charged transition state. Now the solvation of the transition state can approach $\Delta G_{Q\mu}$. This effect, which is described schematically in Fig. 9.7, resembles to some extent the process of using chemical bonding to close a ring and to form a molecule that provides an effective binding site for ions. Thus, we may view enzyme-active sites as *"super solvents"* that provide optimal solvation for the transition states of their reacting fragments (Refs. 10a and 17). As indicated above, this requires a *very polar environment with small reorganization energy* (which may also be described as fixed permanent dipoles in a relatively nonpolar environment, Ref. 10a). This description is the exact opposite from viewing or modeling enzyme-active sites at low dielectric environments that provide small reorganization energies (Ref. 8), since such

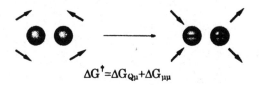

$$\Delta G^{\dagger} = \Delta G_{Q\mu} + \Delta G_{\mu\mu}$$

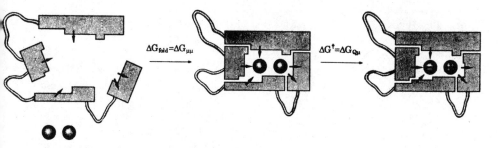

$$\Delta G_{\text{fold}} = \Delta G_{\mu\mu} \qquad \Delta G^{\dagger} = \Delta G_{Q\mu}$$

FIGURE 9.7. A demonstration of the relationship between folding free energy and catalytic energy. The energy balance involved in the formation of an ion-pair type transition state in solution (top). The corresponding energetics in proteins. $\Delta G_{\mu\mu}$ designates the dipole–dipole interaction of the solvent (bottom).

sites would lead to *large* rather than small activation barriers due to their desolvation effect (see Section 9.2.2 and Ref. 17).

In view of the arguments presented in this chapter, as well as in previous chapters, it seems that electrostatic effects are the most important factors in enzyme catalysis. Entropic factors might also be important in some cases but cannot contribute to the increase of k_{cat}/K_M. Furthermore, as much as the correlation between structure and catalysis is concerned, it seems that the complimentarity between the electrostatic potential of the enzyme and the change in charges during the reaction will remain the best correlator. Finally, even in cases where the source of the catalytic activity of a given enzyme is hard to elucidate, it is expected that the methods presented in this book will provide the crucial ability to examine different hypothesis in a reliable way.

REFERENCES

1. L. Pauling, *Chem. Eng. News*, **263**, 294 (1946).
2. T. C. Bruice, *Ann. Rev. Biochem.*, **45**, 331 (1976).
3. D. R. Storm and D. E. Koshland, *J. Am. Chem. Soc.*, **94**, 5805 (1972).
4. M. I. Page and W. P. Jencks, *Proc. Natl. Acad. Sci. U.S.A.*, **68**, 1678 (1971).
5. W. P. Jencks, *Catalysis in Chemistry and Enzymology*, Dover Publication, New York, 1986.
6. P. F. Menger, *Acc. Chem. Res.*, **18**, 128 (1985).
7. (a) M. J. S. Dewar and D. M. Storch, *Proc. Natl. Acad. Sci. U.S.A.*, **82**, 2225 (1985) (b) R. Wolfenden, *Science*, **222**, 1087 (1983). (c) S. J. Weiner, U. C. Singh, and P. A. Kollman, *J. Am. Chem. Soc.*, **107**, 2219 (1985). (d) S. G. Cohen, V. M. Vaidya, and R. M. Schultz, *Proc. Natl. Acad. Sci. U.S.A.*, **66**, 249 (1970). (e) J. Crosby, R. Stone, and G. E. Lienhard, *J. Am. Chem. Soc.*, **92**, 2891 (1970).
8. L. I. Krishtalik, *J. Theor. Biol.*, **88**, 757 (1980).
9. (a) G. Careri, P. Fasella, and E. Gratton, *Ann. Rev. Biophys. Bioeng.*, **8**, 69 (1979). (b) B. Gavish and M. M. Werber, *Biochemistry*, **18**, 1269 (1979).
10. (a) A. Warshel, *Proc. Natl. Acad. Sci. U.S.A.*, **75**, 5250 (1978). (b) A. Warshel, *Acc. Chem. Res.*, **14**, 284 (1981).
11. J-K. Hwang, G. King, S. Creighton, and A. Warshel, *J. Am. Chem. Soc.*, **110**, 5297 (1988).
12. M. F. Perutz, *Science*, **201**, 1187 (1978).
13. A. Warshel, J. Aqvist, and S. Creighton, *Proc. Natl. Acad. Sci. U.S.A.*, **86**, 5820 (1989).
14. R. T. Raines, E. L. Sutton, D. R. Straus, W. Gilbert, and J. R. Knowles, *Biochemistry*, **25**, 7142 (1986).
15. G. van der Zwan and J. T. Hynes, *J. Chem. Phys.*, **78**, 4174 (1983).
16. D. F. Calef and P. G. Wolynes, *J. Phys. Chem.*, **87**, 3400 (1983).
17. A. Yadav, R. M. Jackson, J. J. Holbrook and A. Warshel, *J. Am. Chem. Soc.* **113**, 4800 (1991).

INDEX

Numbers set in **boldface** indicate pages on which a figure or a table appears.